W0230177

Atmospheric Aerosols

Atmospheric Aerosols

Editor

Rekha Kale

Atmospheric Aerosols

Edited by **Rekha Kale**

Printed in 2017

ISBN: 978-1-68117-132-6

Library of Congress Control Number: 2015936553

© 2016 by
SCITUS Academics LLC,
616, Corporate Way, Suite 2, 4766,
Valley Cottage, NY 10989

www.scitusacademics.com

This book contains information obtained from highly regarded resources. Copyright for individual articles remains with the authors as indicated. All chapters are distributed under the terms of the Creative Commons Attribution License, which permits unrestricted use, distribution, and reproduction in any medium, provided the original author and source are credited.

Notice

Reasonable efforts have been made to publish reliable data and views articulated in the chapters are those of the individual contributors, and not necessarily those of the editors or publishers. Editors or publishers are not responsible for the accuracy of the information in the published chapters or consequences of their use. The publisher believes no responsibility for any damage or grievance to the persons or property arising out of the use of any materials, instructions, methods or thoughts in the book. The editors and the publisher have attempted to trace the copyright holders of all material reproduced in this publication and apologize to copyright holders if permission has not been obtained. If any copyright holder has not been acknowledged, please write to us so we may rectify.

Contents

Preface

Atmospheric Aerosols is a vital problem in current environmental research due to its importance in atmospheric optics, energetics, radiative transfer studies, chemistry, climate, biology and public health. Aerosols can influence the energy balance of the terrestrial atmosphere, the hydrological cycle, atmospheric dynamics and monsoon circulations. Because of the heterogeneous aerosol field with large spatial and temporal variability and reduction in uncertainties in aerosol quantification is a challenging task in atmospheric sciences. Keeping this in view the present study aims to assess the impact of aerosols on coastal Indian station Visakhapatnam and the adjoining Bay of Bengal. An aerosol is a colloid of fine solid particles or liquid droplets, in air or another gas. Aerosols can be natural or not. Examples of natural aerosols are fog, forest exudates and geyser steam.

Editor

Modeling in Earth system science up to and beyond IPCC AR5

Tomohiro Hajima[1], Michio Kawamiya[1], Michio Watanabe[1],
Etsushi Kato[2, 3], Kaoru Tachiiri[1], Masahiro Sugiyama[4],
Shingo Watanabe[1], Hideki Okajima[1], and Akinori Ito[1]

[1]Yokohama Institute for Earth Sciences, Japan Agency for Marine-Earth Science and Technology, 3173-25 Showa-machi, Kanazawa-ku, Yokohama 236-0001, Japan

[2]Center for Global Environmental Research, National Institute for Environmental Studies, 16-2 Onogawa, Tsukuba 305-8506, Ibaraki, Japan

[3]The Institute of Applied Energy (IAE), Shimbashi SY Bldg., 1-14-2 Nishi-Shimbashi 1-Chome, Minato-ku, Tokyo 105-0003, Japan

[4]Policy Alternatives Research Institute, University of Tokyo, 7-3-1, Hongo, Bunkyo-ku, Tokyo 113-0033, Japan

ABSTRACT

Changes in the natural environment that are the result of human activities are becoming evident. Since these changes are interrelated and cannot be investigated without interdisciplinary collaboration between scientific fields, Earth system science (ESS) is required to provide a framework for recognizing anew the Earth system as one composed of its interacting subsystems. The concept of ESS has been partially realized by Earth system models (ESMs). In this paper, we focus on modeling in ESS, review related findings mainly from the latest assessment report of the Intergovernmental Panel on Climate Change, and introduce tasks under discussion for the next phases of the following areas of science: the global nitrogen cycle, ocean acidification, land-use and land-cover change, ESMs of intermediate complexity, climate geoengineering, ocean CO_2 uptake, and deposition of bioavailable iron in marine ecosystems. Since responding to global change is a pressing mission in Earth science, modeling will continue to contribute to the cooperative growth of diversifying disciplines and expanding ESS, because modeling connects traditional disciplines through explicit interaction between them.

REVIEW

Introduction

Changes in the natural environment that are a result of human activities are becoming evident. Although one of the clearest examples is global warming, the problem goes well beyond a single issue and includes ocean acidification and perturbations of the global nitrogen cycle from industrial fixation. These issues are interrelated, and no single one can be addressed without interdisciplinary collaboration between scientific fields such as meteorology, oceanography, geochemistry, biology, and even social sciences. Recognizing the situation, scientists have been debating the necessity of 'Earth system science' (ESS), in which the global environment is recognized as a system composed of its interacting subsystems - the atmosphere, oceans, biosphere, cryosphere, and society.

Based on preceding studies on the concept of ESS, a report by the NASA Advisory Council ([1988]) first used the term 'Earth system science' explicitly and provided the clear definition used today. The report set the goal of ESS as the scientific understanding of the entire Earth system on a global scale by describing how its component parts and their interactions have evolved, how they function, and how they may be expected to continue to evolve on all time scales. The report points out that accomplishing this goal will require various research schemes such as numerical modeling, global observation systems, and information networks that enable efficient dissemination of observed data and research outputs. The statements made showed surprising foresight considering that at the time of publication, the use of the Internet was limited to certain advanced institutes and there were very few attempts to incorporate biogeochemical processes into general circulation models (GCMs).

Models used in Earth system science

As the aforementioned report emphasized, modeling can be a powerful tool for investigating the dynamics of the Earth system. Models that have been developed and applied for this purpose can be categorized by their degree of complexity and integration. Models in the 'conceptual' category, which are the least complex, consist of several simple equations mimicking certain aspects of complex behaviors of the Earth system (e.g., Budyko [1969]; Sellers [1969]). The degree of abstraction is, however, extremely high for this type of model, and correspondence between model equations and processes in nature is not readily understood. This leads to a fundamental difficulty in estimating model parameter values. Models in this category are therefore mainly used for educational purposes or as supporting material for constructing a theoretical framework and are rarely used for projection.

In contrast, atmospheric and oceanic GCMs have been applied to the projection of El Niño events, global warming, and others. GCM-based Earth system models (ESMs), which are introduced in the next section, have a drawback in that they are computationally expensive. To fill the gap between conceptual models and GCMs, ESMs of intermediate complexity (EMICs) are now being extensively developed (Claussen et al. [2002]). EMICs greatly simplify equations of motion for the atmosphere and ocean and radiation processes, retaining some ability to reproduce realistic properties such as geographic temperature distribution and deep water formation. EMICs require much less in terms of computer resources and can be integrated for many thousands of years without supercomputers.

With careful experiment design, EMICs constitute an important element for future projection and interpretation of past events with time scales longer than approximately a few hundred years (e.g., Timmermann et al. [2009]; Hargreaves et al. [2012]). However, the same drawback of substantial abstraction for conceptual models applies to EMICs, albeit to a lesser extent. It is desirable to make parallel and complementary use of GCM-based ESMs and EMICs within computer resource availability.

Many institutes in the world involved in global warming projection are developing elaborate ESMs by coupling biogeochemical modules (vegetation dynamics, the carbon cycle, and others) with GCMs (Figure 1). This figure shows components specific to the Model for Interdisciplinary Research on Climate, Earth System Model (MIROC-ESM; Watanabe et al. [2011]). This is a GCM-based ESM developed by the Japan Agency for Marine-Earth Science and Technology (JAMSTEC), with collaboration from The University of Tokyo Atmosphere Ocean Research Institute (AORI) and National Institute for Environmental Studies (NIES). Most ESMs developed by leading institutes have similar structures. One of the most striking findings from studies using ESMs is that there is positive feedback between climate change and the carbon cycle (IPCC [2013]), mainly caused by response of the terrestrial biosphere to rising temperature.

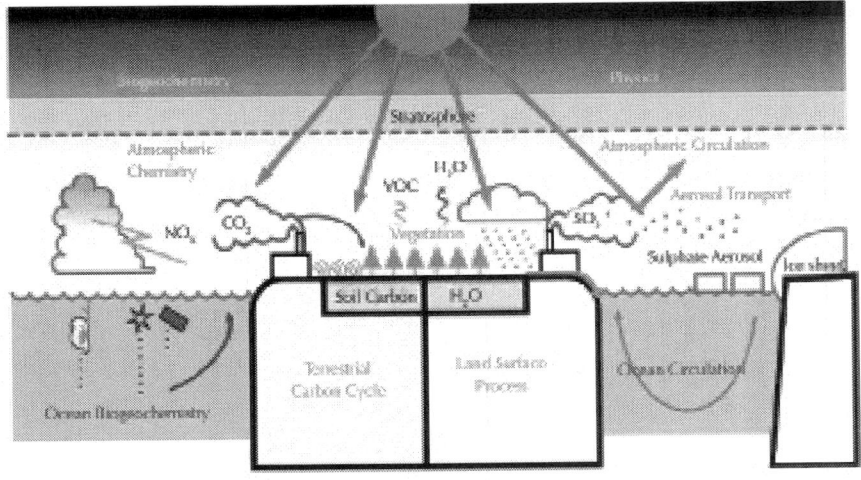

Figure 1: Components in MIROC-ESM. The MIROC-ESM is a GCM-based earth system model developed jointly by the Japan Agency for Marine-Earth Science and Technology, Center for Climate System Research of the University of Tokyo, and the National Institute for Environmental Studies.

For coupling the carbon cycle to climate models, a new method of climate-carbon cycle simulation has been devised and adopted as a common method of climate-carbon cycle projection. Traditionally, ESMs are driven by 'CO$_2$ emission', in which future climate change is projected by internally predicted CO$_2$ concentration based on prescribed anthropogenic CO$_2$ emission (e.g., Friedlingstein et al.[2006]; Yoshikawa et al. [2008]; Friedlingstein et al. [2014]). This type of simulation reflects the actual sequence of the global carbon cycle and is therefore very intuitive. However, it is difficult to effectively compare ESM results with other conventional GCMs, because different CO$_2$ concentrations are referenced in these two types of climate model.

Alternatively, in the new method, with so-called 'concentration-driven' simulations, the rates of the atmosphere-ocean and atmosphere-land CO$_2$ exchanges are diagnosed in ESMs by referring to the prescribed CO$_2$ concentration and predicted climate. Because uncertainty arising from carbon cycle processes can be excluded in the projected climate, we can effectively compare climate-related variables (such as temperature and climate sensitivities) between ESMs and GCMs (e.g., Andrews et al. [2012]; Knutti et al. [2013]).

Another feature of this concentration-driven method is that by using the prescribed CO_2 concentration and diagnosed CO_2 fluxes, we can inversely estimate the anthropogenic CO_2 emission that is allowed to achieve the prescribed CO_2 concentration pathways. Since the rate of emission is almost equivalent to that of anthropogenic fossil fuel CO_2, this inversely estimated CO_2 emission in the experiments is sometimes called 'allowable' or 'compatible' emission. For example, in the study of Jones et al. ([2013]), compatible emissions simulated by ESMs were compared by applying prescribed CO_2 pathways of representative concentration pathways (RCPs). These authors found that about half the participating models predicted that 'negative' anthropogenic emissions will be necessary during the twenty-first century to realize RCP2.6 scenarios. As such information from the concentration-driven experiments has a direct implication for future climate mitigation policies, this simulation style is now becoming popular among those who use ESMs, suggesting a new application of models in ESS for global environmental change projections.

Future Directions

It has been pointed out that the behavior of the simulated terrestrial biosphere may be drastically changed if one includes the nitrogen cycle. The intensity of climate-carbon cycle feedback based on models with a global nitrogen cycle will be an issue of great interest for the possible next phase of the Intergovernmental Panel on Climate Change (IPCC) assessment, as explained in the 'Nitrogen cycle' section.

Although the ocean exhibits feedback caused by the temperature dependence of the CO_2 solubility in seawater, its intensity is about a quarter of that of the terrestrial biosphere, at least under the current setup of typical ESMs, often without a terrestrial nitrogen cycle (IPCC [2013]). The projection of the ocean uptake of CO_2 is nevertheless critical, because the ocean will be the primary sink of anthropogenic CO_2 (details in 'Ocean CO_2 uptake' section) and the uptake causes another problem, ocean acidification caused by the weak acidity of CO_2. The ocean acidification problem is reviewed in more detail in the 'Ocean acidification' section.

It is sometimes purported that the role of global environmental projection is gradually changing from science used as a warning to that which can be implemented (Kerr [2011]). Indeed, 'actionable science' was the watchword in the 2011 Open Science Conference of the World Climate Research Programme (WCRP) in Denver (Asrar et al. [2012]). Climate models can serve society in many ways, such as providing information on extreme weather. ESMs are particularly useful for reflecting scientific perceptions of future socioeconomic scenarios. Components of ESMs for global cycles of carbon and other biogeochemical properties can act as a bridge between socioeconomic models, whose outputs are frequently greenhouse gas emissions, and climate models, which require concentrations of the gases. Another aspect of the way in which ESMs are helpful for linking climate science and the socioeconomy is that they can deal with land-use change (LUC) processes, a key factor for the future global carbon budget and regional climate change. Issues of LUC are surveyed in the 'Land-use and land-cover change' section. Care must be taken regarding uncertainties when ESM results are applied to developing socioeconomic scenarios. Uncertainties in parameter values of ESMs are even more serious than those in conventional climate models, because ESMs incorporate biological processes such as photosynthesis. Parameter ensemble experiments are desirable but not always feasible, owing to the computational cost of GCM-based ESMs. Using GCM-based ESMs and EMICs in parallel may remedy this problem, as discussed in the section titled 'Earth system models of intermediate complexity'.

Actionable science might go well beyond just providing information. Artificial, active modification of the global climate, often termed geoengineering, is now a matter of debate. The Fifth Assessment Report (AR5) of IPCC (IPCC [2013]) treated this issue in multiple chapters. This should, however, be regarded as a last resort since most of its side effects are yet to be assessed and there will undoubtedly be many others of which we are not yet aware. Scientists have started a project to evaluate the effects of geoengineering using numerical models. Results from that project will be introduced in the section titled 'Climate geoengineering,' together with other aspects of geoengineering presented in the 'Ocean CO_2 uptake' and 'Atmospheric deposition of bioavailable iron in marine ecosystems' sections.

This review is not intended to cover all subjects that can be tackled with ESMs. Paleoclimate is an example of topics missing here, but this does not mean that the authors consider paleoclimate to be unimportant. Rather, collaboration between communities of paleoclimate and global environmental change projection is critical to enhancing the credibility of future projection, since paleoclimate provides the only means for validating climate models on time scales longer than decades. A limited number of topics are addressed in this review, from an extremely large number of fields in which ESMs can be applied.

Nitrogen Cycle

Modern Earth system modeling may be said to have begun by incorporating land/ocean carbon cycle processes into GCMs (Cox et al. [2000]). The latest ESMs are now equipped with more complex and interactive processes on global chemical/biogeochemical processes, such as atmospheric chemistry. Recent studies analyzing simulation results from the Coupled Model Intercomparison Project phase 5 (CMIP5) ESMs suggest the importance of the global nitrogen cycle, particularly the process of nitrogen shortage and its limitation on ecosystem productivity (Arora et al. [2013]; Hajima et al.[2014]). This deficiency of available nitrogen could alter the behavior of the global carbon cycle, thereby impacting the global climate. In addition, from the beginning of the twentieth century, the global nitrogen cycle has been greatly perturbed by the human creation of reactive nitrogen and anthropogenic emission of large amounts of N_2O, one of the strongest greenhouse gases (Gruber and Galloway [2008]). To quantify historical human effects on the global nitrogen cycle and make projections for future climate change, ESMs with an explicit nitrogen cycle can be a powerful tool. Here, we first summarize the interaction of climate and the carbon cycle and then describe the impacts of the global nitrogen cycle on the climate by regulating the global carbon budget. Finally, the direct impacts of the nitrogen cycle on the climate due to the production of greenhouse and associated gases are briefly summarized.

Feedback on Carbon Cycle from Environmental Changes

The global carbon cycle has long been regarded as one of the major components that control the global climate because CO_2 in the atmosphere has a greenhouse effect and its concentration is regulated by ocean and land ecosystems through the exchange of CO_2 with the atmosphere. Furthermore, human activities such as fossil fuel combustion, land-use change, and cement production have released large amounts of CO_2 into the air. Its concentration is now reaching 400 ppmv, which corresponds to about a 40 % increase from the preindustrial state. Changes of global carbon balance in the industrial era can be simply understood by

$$FF = \Delta C_A + \Delta C_L + \Delta C_O, \tag{1}$$

Where FF is the cumulative amount of carbon emitted by fossil fuel combustion and ΔC represents the changes in the carbon amounts from the preindustrial state of the atmosphere (A), ocean (O), and land (L). In this formulation, the partitioning of emitted carbon between the atmosphere, ocean, and land ecosystems is variable; environmental changes such as global warming can change the capability of carbon uptake by land and the ocean and thereby alter the airborne fraction of emitted CO_2 (Le Quéré et al. [2013]; Jones et al. [2013] for treatment of land-use change effects on the global carbon balance). For example, global warming could induce large ecosystem respiration and thus reduce total terrestrial carbon. To understand such interactive behavior of the carbon cycle and climate, Friedlingstein et al. ([2003]) proposed a simple mathematical expression to link carbon balance and environmental changes:

$$\Delta C_{L,O} = \beta_{L,O} \Delta C_A + \gamma_{L,O} \Delta T \tag{2}$$

The first term on the right represents a change in the amount of stored carbon due to an atmospheric CO_2 increase, assuming a linear response of carbon storage to increasing atmospheric CO_2 (ΔC_A) with coefficient β. This environmental effect on carbon storage is sometimes called 'CO_2-carbon feedback'. Because an increased CO_2 concentration promotes carbon uptake by land and oceans (by stimulating photosynthesis in land ecosystems and increasing the CO_2 partial pressure deficit between the atmosphere and oceans), β is considered to have a positive sign (i.e., CO_2-carbon feedback is negative) (Arora et al. [2013]; Friedlingstein et al. [2006]; Gregory et al. [2009]). The second term represents a carbon storage change in response to climate change (represented by ΔT, the degree of warming); this process is called 'climate-carbon feedback'. Since warming could cause larger ecosystem respiration and a reduction of CO_2 dissolution in water, the feedback parameter γ is regarded to have a negative sign (Arora et al. [2013]; Friedlingstein et al. [2006]; Gregory et al.[2009]), meaning that interaction between climate and the carbon cycle forms a positive feedback loop.

By combining these two equations, we obtain an expression that incorporates environmental change (ΔC_A and ΔT) into the global carbon balance:

$$FF = \Delta C_A + \beta_{L,O} \Delta C_A + \gamma_{L,O} \Delta T \qquad (3)$$

This equation indicates that cumulative carbon contained in fossil fuel emission increases atmospheric carbon (ΔC_A), but carbon partitioning in the atmosphere depends on the land/ocean carbon cycle response, CO_2-carbon ($\beta \Delta C_A$) and climate-carbon ($\gamma \Delta T$) feedbacks. A recent study by Arora et al. ([2013]) in CMIP5 compared these feedback strengths for nine ESMs via sensitivity analysis (Table 1), showing that land and ocean carbon cycles have comparable levels of CO_2-carbon feedback. As multi-model means, β was 0.92 PgC ppmv^{-1} for land and 0.80 for the ocean. The climate-carbon feedback for land was greater (negative) than that for the ocean, at -58.4 and -7.8 PgC K^{-1} respectively. In their simulations, although the CO_2 growth rate was rapid (CO_2 increase of 1.0% year^{-1} during 140 simulation years) and thus substantial global warming was expected, most ESMs showed that CO_2-carbon feedback surpassed climate-carbon feedback across all simulation periods, so natural carbon sinks continued to absorb anthropogenic carbon in the simulations.

Table 1: Equilibrium climate sensitivity (ECS), and concentration (β) and climate (γ) carbon cycle feedback strengths

	EMICs (8 models)	CMIP5 ESMs (15 models for ECS, 9 models for others)
ECS (K)	3.04 ± 0.67	3.37 ± 0.83
βL (PgC/ppm)	0.69 ± 0.31	0.92 ± 0.44
βo (PgC/ppm)	0.93 ± 0.19	0.80 ± 0.07
γL (PgC/K)	−61.5 ± 29.7	−58.4 ± 28.5
Γ (PgC/K)	−9.6 ± 6.7	−7.8 ± 2.9

These parameters were estimated by EMICs and ESMs. Variables were obtained from Eby et al. ([2013]) for EMICs and from Andrews et al. ([2012]) and Arora et al. ([2013]) for ESMs. L, land; O, ocean.

Hajima et al.

Hajima et al. Progress in Earth and Planetary Science 2014 1:29, doi: 10.1186/s40645-014-0029-y

Nitrogen Cycle Feedback on Carbon Cycle

The global nitrogen cycle is one of the most important biogeochemical cycles, together with that for carbon. Although the total mass flux of global nitrogen is much smaller than that of carbon (Gruber and Galloway [2008]), the nitrogen cycle may have strong impacts on the carbon cycle. Since nitrogen is used by organisms to make amino acids, enzymes, proteins, and nucleic acids, the nitrogen cycle in ecosystems is intimately associated with the life functions of organisms. It is used to develop organisms' bodies, maintain their activity, and increase their populations (Canfield et al.[2010]). However, since most nitrogen in nature is in an inactive form (N_2), nitrogen in reactive and available forms for ecosystems (such as ammonia and nitrous oxide) may be rare and thus often be a limiting factor for ecosystem productivity. This deficiency of nitrogen can strongly restrict carbon fluxes in ecosystems, hence having a feedback effect on the global climate.

In Equation 3, the nitrogen cycle feedback on the climate-carbon cycle system can be summarized as follows. First, the amount of available nitrogen for plants and phytoplankton can control their productivity and the resultant carbon fluxes out of the atmosphere to land or ocean. As shown by free-air CO_2 enrichment studies, an elevated CO_2 concentration on land can stimulate photosynthesis and accumulate biomass. However, if the amount of available nitrogen does not meet demands for plant growth, the plant biomass accumulation rate is likely constrained (e.g., Reich et al. [2006]; Reich and Hobbie [2013]). For ocean ecosystems, the gross productivity of phytoplankton and its spatial distribution are originally and strongly regulated by the available nitrogen amount, even in current conditions (Canfield et al. [2010]). Because most terrestrial ecosystem models in ESMs are now incapable of explicitly representing the influence of a nitrogen deficit on photosynthetic capacity or other physiological aspects (e.g., leaf area), nitrogen feedback should be reflected by the parameter β to weaken the strength of negative CO_2-carbon feedback. In fact, some CMIP5 ESMs incorporating explicit terrestrial nitrogen cycle processes show a weaker response to increasing CO_2 than those output by other models, which is likely due to the nitrogen-limited response of plant productivity in an enriched CO_2 world (Arora et al. [2013]; Hajima et al. [2014]).

Furthermore, the nitrogen cycle on land could affect the carbon storage response to climate change, as represented by parameter γ in Equation 3. Since global warming could lead to enhanced activities of soil microbes, accelerated soil decomposition may reduce the total amount of soil carbon. However, accompanied by the release of inorganic nitrogen that becomes soil nutrients, this accelerated soil decomposition rate might activate plant productivity. In fact, some models incorporating explicit terrestrial nitrogen cycle processes reduce the carbon cycle response to global warming, with less negative or sometimes slightly positive values for (Sokolov et al. [2008]; Bonan and Levis [2010]; Thornton et al. [2009]; Zaehle et al. [2010]). In these models, although global warming reduces soil carbon storage because of enhanced heterotrophic respiration, increasing soil nutrients somewhat compensates the global terrestrial carbon reduction by increasing vegetation carbon storage.

The global nitrogen cycle could alter the global climate by changing both the CO_2-carbon and climate-carbon feedback, thereby creating climate-carbon-nitrogen interactions. Although the number of studies is limited, ESMs with a nitrogen cycle show a similar trend in that the incorporation of a nitrogen cycle reduces the sensitivity of the carbon cycle response to environmental changes. However, constraints of the nitrogen cycle on the carbon cycle should be addressed with the presence of reactive nitrogen in the atmosphere, as described below.

Direct Impact of the Nitrogen Cycle on Climate and Human Perturbations

In the previous section, we described nitrogen cycle feedbacks on the climate through the regulation of carbon cycle responses. In addition, the nitrogen cycle directly impacts global climate because the greenhouse gas N_2O has a relatively long lifetime in the atmosphere, more than 100 years. Galloway et al. ([2004]) estimated the total N_2O emission under conditions before the significant human perturbation at approximately 12 TgN year^{-1}. Its emission rate has greatly increased during the industrial era because human activities such as industrial nitrogen fixation for fertilizer, fossil fuel/biomass combustion, and agriculture have increased the total amount of reactive nitrogen globally (Galloway et al. [2004]; Gruber and Galloway [2008]). Some studies using data assimilation or inversion techniques established the current total global N_2O emission at approximately 18 TgN year^{-1}, and the contribution of the anthropogenic emission to the global total at approximately one third (Saikawa et al. [2013] estimate for 2002 to 2005). The latter emission has increased its concentration in the atmosphere and thus N_2O radiative forcing is now considered to be 0.17 W m^{-2}, about 10% of CO_2 radiative forcing (IPCC [2013]). Furthermore, such human activities have created other forms of reactive nitrogen, namely, NOx, NHx, and NOy. Nitrogen in these reactive forms interacts with global and local climates by contributing to the formation of aerosols, acting as a precursor for generating tropospheric ozone, and through its association with CH_4 reduction (Menon et al. [2007]).

Reactive nitrogen in these forms may further affect carbon-nitrogen interactions. Since net radiative forcing of these agents is positive

(IPCC [2013]), there could be an additional warming that could accelerate soil decomposition and modify carbon cycle feedbacks on the climate. In addition, since inorganic mineral nitrogen could be an ecosystem resource for productivity, its deposition on land and ocean surfaces could also have indirect impacts on the climate by alleviating the nitrogen limitation on productivity (e.g., Bonan and Levis [2010]; Thornton et al. [2009]). Although it is important to assess the combined effect of direct (changing atmospheric composition of non-CO_2 greenhouse gases) and indirect (changing CO_2 concentration via modulating carbon cycle feedbacks) impacts of the nitrogen cycle on the global climate, the number of related studies is limited (e.g., Stocker et al.[2013]; Zaehle et al. [2011]). For comprehensive understanding of the influence of human activity on the global nitrogen cycle and its propagation impacts on the global environment, it is hoped that more scientific efforts will be made using fully coupled carbon-nitrogen-climate models.

Ocean Acidification

Definition of Ocean Acidification

The emission of large amounts of anthropogenic carbon dioxide has increased the global atmospheric CO_2 concentration and contributes to temperature increases in the atmosphere and ocean. Since 1750, the global ocean has absorbed about a third of anthropogenic CO_2 released into the atmosphere (Sabine et al. [2004]; Sabine and Feely [2007]). CO_2 reacts with water molecules (H_2O) to form the weak acid H_2CO_3 (carbonic acid), and most of this acid dissociates into hydrogen ions (H^+) and bicarbonate ions (HCO_3^-), such as

$$H_2O + CO_2 \leftrightarrow H_2CO_3$$

$$H_2CO_3 \leftrightarrow H^+ + HCO_3^-$$

Some of the resulting H⁺ reacts with carbonate ions CO_3^{2-} to produce additional HCO_3^- ions.

$$H^+ + CO_3^{2-} \leftrightarrow HCO_3^-$$

As a result, CO_2 dissolution in the ocean increases H⁺ (thereby decreasing pH) and reduces CO_3^{2-} concentrations, a process known as ocean acidification (Broecker and Clark [2001]; Caldeira and Wickett [2003], [2005]; Doney et al. [2009]). Figure 2 shows a time series of the atmospheric CO_2 at Mauna Loa as well as the surface ocean CO_2 partial pressure (pCO_2) and surface ocean pH at the ocean station ALOHA in the subtropical North Pacific Ocean. We see that as atmospheric CO_2 rises, some extra CO_2 is transferred into ocean surface waters, leading to ocean acidification. The pH of ocean surface waters has decreased by about 0.1 since the dawn of the industrial era (Caldeira and Wickett [2003], [2005]), with a decrease of approximately 0.0018 year⁻¹ observed over the last quarter century at several open-ocean time-series sites (Bates [2007]; Bates and Peters [2007]; Santana-Casiano et al. [2007]; Dore et al. [2009]).

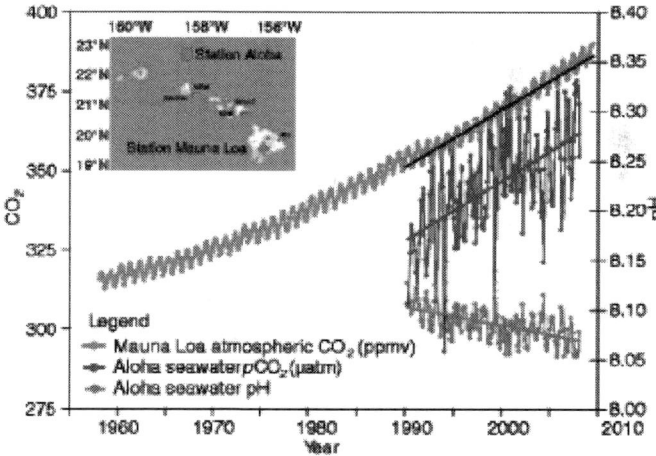

Figure 2: Observed partial CO_2 pressure (pCO_2) and ocean pH at ocean surface. pCO_2 (µatm; blue) and surface ocean pH (green) at ocean are observed at Station ALOHA in subtropical North Pacific Ocean, in

comparison with atmospheric CO_2 concentration observed at Mauna Loa (in parts per million by volume; red). Note that the increase in oceanic CO_2 over the past 17 years is consistent with the atmospheric increase, within the statistical limits of the measurements (Doney et al. [2009]).

Marine calcifying organisms such as plankton, shellfish, coral, and fish use carbonate ions CO_3^{2-} to build their shells or skeletons from calcium carbonate ($CaCO_3$),

$$Ca^{2+} + CO_3^{2-} \leftrightarrow CaCO_3,$$

So they are expected to be greatly affected by ocean acidification. The carbonate ion concentration is often expressed by the degree of saturation of biominerals aragonite (Ω_{Ar}) and calcite (Ω_{Ca}) (Feely et al. [2004]). Shell and skeleton formation generally occurs in supersaturated sea water $\Omega > 1$ and dissolution in undersaturated seawater $\Omega < 1$. Consequently, spatial and temporal changes in saturation state with respect to these mineral phases are important for understanding how ocean acidification might substantially impact future ecosystems.

Assessment by ESMs

We can assess the ocean's present and future ability to take up anthropogenic CO_2 and the influence of ocean acidification using ESMs in which the ocean carbon cycle is included. Multi-model projections using ocean process-based carbon cycle models have demonstrated large decreases in pH and carbonate ion concentration CO_3^{2-} during the twenty-first century, across all the world's oceans. By the middle of the present century, atmospheric CO_2 levels could reach more than 500 ppm and exceed 800 ppm by the end of the century (Friedlingstein et al. [2006], [2014]). By 2100, these CO_2 levels would result in an additional decrease of surface water pH by 0.3 units over current conditions and 0.4 over the preindustrial level. This represents an increase in the ocean's hydrogen ion concentration H^+ by 2.5 times relative to the beginning of the industrial era (Orr et al. [2005]; Fe onite. ESMs show that this aragonite undersaturation in surface waters is reached before the end of the twenty-first century in the Southern Ocean (Orr et al. [2005]). Steinacher et al. ([2009]) suggested that aragonite undersaturation occurs sooner and is more intense in the Arctic. When atmospheric CO_2 reaches 428 ppm, 10% of Arctic surface waters are projected to become undersaturated. By 2100, atmospheric CO_2 will exceed 800 ppm and much of the Arctic surface is projected to become undersaturated with respect to calcite (Feely et al. [2009]). This means that surface waters would be corrosive to all $CaCO_3$ minerals.

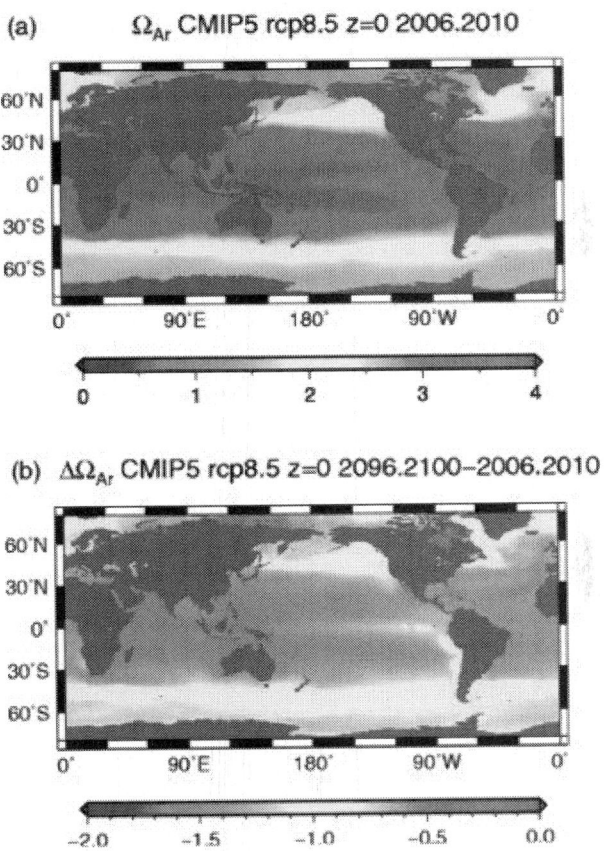

Figure 3: Spatial distribution of aragonite saturation state ΩAr.(a)Model-mean Ω_{Ar} at the sea surface averaged over 2006 to 2010, derived from nine CMIP5 ESMs under the AR5 RCP8.5 scenario. (b) Change in Ω_{Ar} at the sea surface from 2006-2010 to 2096-2100.

These changes extend well below the sea surface (Orr et al. [2005]). Throughout the Southern Ocean, the entire water column becomes undersaturated with respect to aragonite. During the twenty-first century, the aragonite saturation horizon, representing the limit between undersaturation and supersaturation in the Southern Ocean and subarctic Pacific, shoals to the sea surface. In the North Atlantic, surface waters remain saturated with respect to aragonite, but the aragonite saturation horizon shoals dramatically, e.g., north of 50°N it shoals from 2,600 to 115 m. Greater erosion in the North Atlantic is due to deeper penetration of waters with higher concentrations of anthropogenic CO_2 from the sea surface (Sarmiento et al. [1992]; Gruber [1998]; Sabine et al.[2004]).

Surface $CaCO_3$ saturation also varies seasonally, particularly at high latitudes, where observed saturation is higher in summer and lower in winter (Merico et al. [2006]; Findlay et al. [2008]). Future projections using ocean carbon cycle models indicate that undersaturated conditions will be reached first in winter (Orr et al. [2005]). In the Southern Ocean, it is projected that wintertime undersaturation with respect to aragonite will begin when atmospheric CO_2 reaches 450 ppm, which is about 100 ppm sooner (30 years) than for the annual mean undersaturation (McNeil and Matear[2008]). Aragonite undersaturation will be reached during wintertime in parts (10%) of the Arctic when atmospheric CO_2 attains 410 ppm (Steinacher et al. [2009]).

Controlling Factors

As mentioned above, future reductions in surface ocean pH and $CaCO_3$ (calcite and aragonite) saturation states are mostly controlled by the uptake of anthropogenic CO_2. Other effects of future climate change counteract less than 10% of the reductions of $CaCO_3$ saturation (Orr et al. [2005]; McNeil and Matear [2008]). An exception is the Arctic Ocean, where decreases in the pH and aragonite saturation state are predicted to be caused by increased freshwater input from sea ice melt, enhanced precipitation, and stronger air-sea CO_2 fluxes because of less sea ice cover (Steinacher et al. [2009]; Yamamoto et al. [2012]). This result indicates that future projections of the pH and aragonite saturation state at high latitudes may be significantly influenced by rapid sea ice reduction as well as increases of atmospheric CO_2 concentration.

Focusing on the regional ocean carbon cycle, models project some nearshore systems to be highly vulnerable to future pH decrease. In the California Current system, an eastern upwelling system, strong seasonal upwelling of carbon-rich waters (Feely et al. [2008]) makes surface waters sensitive to future ocean acidification, as in the Southern Ocean. In the northwestern European shelf seas, large spatiotemporal pH variability is enhanced by local effects from river input and organic matter degradation, exacerbating ocean acidification from anthropogenic CO_2 invasion (Artioli et al. [2012]). In the Gulf of Mexico and East China Sea, coastal eutrophication, another anthropogenic perturbation, has been shown to enhance subsurface acidification as additional respired carbon accumulates at depth (Cai et al. [2011]).

Land-Use and Land-Cover Change

On the land surface, about one-third to one-half the area of natural ecosystems has been converted to managed land for use as cropland and pasture over the past 10,000 years (Klein Goldewijk et al.[2011]). Human land use is projected to expand in the future because of food and energy demands due to changes in the population and socioeconomic factors (Bruinsma [2009]). Anthropogenic modification of land cover and its management affect ecosystem functioning and also modifies Earth system atmosphere-biosphere interactions through biogeophysical and biogeochemical effects (Claussen et al. [2001]; Pongratz et al. [2010]; Sato et al. [2014]). Changes in surface albedo and latent and sensitive heat fluxes are principal biogeophysical effects of land-use and land-cover changes, which affect atmospheric conditions via changes to the hydrologic cycle and energy balance (Bonan [2008]). A striking example of the biogeophysical effect of land-use change is that of historical deforestation at mid and high latitudes, which has increased surface albedo in winter by altering snow cover and lowering surface air temperature (Davin and de Noblet-Ducoudré [2010]). In the tropics, deforestation reduces evapotranspiration, which also affects atmospheric conditions (Bala et al. [2007]). Furthermore, irrigation and soil management on cropland is known to affect local air temperatures (Lobell et al. [2006]).

The dominant biogeochemical effect of historical land-use and land-cover change is an increase in the atmospheric CO_2 concentration, through decreases in terrestrial carbon stock via expansion of anthropogenic land use (Sato et al. [2014]). About 33% of total anthropogenic CO_2 emissions from the preindustrial era are from land-use change (Houghton et al. [2012]). It is also pointed out in several studies (Bouwman et al. [2005]; Bodirsky et al. [2012]) that increases in atmospheric N_2O, for which the global-warming potential over 100 years is about 300 times that of CO_2, are caused by emissions from the increased use of fertilizer on cropland.

For a quantitative understanding of the biogeophysical and biogeochemical effects of land-use and land-cover change, model intercomparison was first done using EMICs. These couple terrestrial biogeochemical models with simplified lower-resolution climate models (Brovkin et al. [2006]) and then GCMs with land surface schemes, with consideration of anthropogenic land-cover type (Pitman et al. [2009]). To incorporate changes of land-cover type, anthropogenic land use as a fraction of each land-use type for a grid cell in forcing data is assigned annually in the model. To represent biophysical properties of anthropogenic land-use type such as cropland, pasture, and urban (if considered in the land-use data), each plant functional type that has corresponding land surface scheme parameters (such as phenological, morphological, and photosynthetic) is allocated a percentage at a grid cell annually, according to land-use forcing data. Effects on the carbon budget caused by land-cover and land-use change are estimated through carbon emissions from deforestation and wood harvesting. This estimation is done using several product pools with various time scales of decay (McGuire et al. [2001]) and additional carbon uptake after the abandonment of cropland by vegetation regrowth, which can be affected by the age distribution of vegetation on secondary land (Shevliakova et al. [2009]).

Pongratz et al. ([2010]) evaluated the effect of land-use change throughout the twentieth century using a GCM-based ESM. They estimated the effect of land-use change on the global mean temperature in the twentieth century to be −0.03°C through biogeophysical effects and 0.18°C through biogeochemical effects, giving a net effect of 0.15°C. Generally, biogeophysical effects of historical land-use change on the global mean surface air temperature are less than biogeochemical effects, because the former operates locally and biogeophysical changes from deforestation (e.g., those of albedo and latent heat flux) sometimes work together to compensate their individual impacts on temperature (Pitman et al. [2009]; Pongratz et al. [2010]).

Uncertainty of Land-use and Land-cover Change

In such evaluations of impacts on the atmosphere by global land-use change, large uncertainties persist among model estimates (Sato et al. [2014]). For example, in the study of the current global carbon budget, one of the largest uncertainties originates from anthropogenic land-use changes in terrestrial ecosystems. The standard deviation of carbon flux caused by historical anthropogenic land-use change is estimated to be in the order of ±0.5 PgC (Le Quéré et al. [2013]). The choice of land-use data could be one of the causes of estimate variations in land-use change emissions (Jain and Yang [2005]; Meiyappan and Jain [2012]). In the global evaluation of land-use and land-use change effects, land-use data compiled at an approximately 0.5° × 0.5° spatial resolution are generally used. Uncertainty in gridded historical land-use data derives from differences in cropland and pasture area between datasets, downscaling processes of regionally aggregated data, and others. These uncertainties in the reconstruction of historical land-use data should be considered in model evaluation (Klein Goldewijk and Verburg [2013]). In a recent study of the global carbon budget, the standard deviation of estimated land-use change flux with multiple terrestrial carbon cycle models using the same land-use dataset was estimated at 0.42 PgC year^{-1} (Le Quéré et al. [2013]). In contrast, the standard deviation estimated by a single model, which was also used in the multiple model comparison using three different land-use datasets, was 0.27 PgC year^{-1} (Jain et al. [2013]).

Care should be taken regarding whether gross transition information between each land-use category in a grid cell is available within the land-use dataset. This is because some models use net changes and others gross changes of land-use type in the calculation of the carbon budget. In addition, there are several other problems in the use of datasets, which may cause further uncertainties because of a lack of detailed protocol for the handling of these. These include differences in the implementation of land-use data in the terrestrial ecosystem component among models, and whether the model considers emissions and uptake caused by shifting cultivation, wood harvesting, forest degradation, crop harvesting, peat fires, and others.

LUC in CMIP5 and Its Future Direction

In CMIP5, anthropogenic land-use and land-use changes were considered in historical and future scenarios simulated by ESMs. Harmonized land-use transition datasets have been prepared through historical (1500 to 2005) and four future RCPs through 2100 (Hurtt et al. [2011]). The historical part consists of the reconstructed cropland and pasture dataset 'History Database of the Global Environment' (HYDE) (Klein Goldewijk et al. [2010]; Klein Goldewijk et al. [2011]) and wood harvest data of the Food and Agriculture Organization (FAO) of the United Nations. Four future scenarios constructed by integrated assessment models (IAMs) were harmonized with the historical data to smoothly connect past and future scenarios in 2005, using consistent land-use categories. These are primary land, secondary land, cropland, pasture, and urban areas, with a 0.5° × 0.5° horizontal resolution. Annual transitions among these land-use categories, wood harvest area, and carbon mass changes were calculated for each grid cell. In the CMIP5 experiments, ESM modelers used harmonized land-use data as common forcing to consider the effects of land-use change on the terrestrial carbon cycle and biogeophysical effects of land-cover change in climate simulations (Taylor et al. [2012]). In addition to the standard CMIP5 experiments, the international project Land-Use and Climate, Identification of Robust Impacts (LUCID) conducted LUCID-CMIP5 experiments using ESMs, in which land use was fixed at 2005 to evaluate future land-use scenarios (Brovkin et al. [2013]). Under the LUCID framework, biogeophysical effects and variation of the carbon cycle from future land-use change was evaluated using concentration-driven simulations of RCP2.6 and RCP8.5 (Brovkin et al. [2013]; Boysen et al. [2014]). In this simulation, there were no significant impacts on the global climate from biogeophysical changes caused by future land-use change. However, in regions exceeding 10% land-use change during the scenario period, half the models showed significant impacts of such change on surface temperature. Furthermore, some models revealed significant changes in surface albedo, latent heat flux, and available energy in the 10% exceedance region. Nevertheless, these biogeophysical effects on climate in the future scenarios were weak because the area of intense land-use change was restricted to tropical and subtropical regions and the changed area was small compared with historical changes. For effects on terrestrial carbon storage, all models showed significant decreases in carbon stock in the two RCP scenarios. Estimations of cumulative carbon emission caused by land-use change varied greatly among the models, suggesting scientific issues in implementing land-use change processes in the ESMs (e.g., some models considered both crop and pasture areas, whereas others treated only cropland; some models used information on gross land-use transitions, whereas others used net transitions).

After the CMIP5 exercise, needs were recognized for more detailed protocols to handle anthropogenic land use and land-use change in land surface models. Consideration of more precise land-use categories and management processes such as irrigation, wood harvest treatment, biofuel crop type, and afforestation have been addressed for the upcoming 6th phase of Coupled Model Intercomparison Project (CMIP6). This is because the future carbon cycle is strongly dependent on the land-use scenario as well as the terrestrial ecosystem response to both future CO_2 and climate change (Jones et al. [2013]). Toward CMIP6, experiments involving land-use are also considered in specialized intercomparisons that would use CMIP6 standards and infrastructure. In this process, especially for future scenario simulations, the validation of simulated carbon stock in contemporary land ecosystems would be of great importance, because bias in CMIP5 historical simulations is one cause of strong variability in the simulated future changes of carbon stock within CMIP5 and LUCID-CMIP5 simulations (Brovkin et al. [2013]; Anav et al. [2013]). In addition, C-N interaction for CO_2 fertilization effects should be considered because of their strong impacts on vegetation regrowth following land-use change, which in turn affects contemporary and future terrestrial carbon sink trends (Jain et al. [2013]).

Earth System Models of Intermediate Complexity

State-of-the-art GCMs/ESMs furnish irreplaceable future projection data. These models are designed to maximize the representation of detailed processes while retaining reasonable computational speed so that experiments are completed as scheduled, and are unsuitable for executing very long runs or ensemble experiments sufficiently large to extract statistically useful information. In contrast, simple box models aid conceptual understanding, but a lack of geographic detail prevents practical applications and their implications are more qualitative than quantitative. EMICs are designed to fill the gap between the two model types. EMIC results are sometimes included in inter-model comparisons of ESMs, e.g., the Coupled Climate Carbon Cycle Model Intercomparison Project (C4MIP) (Friedlingstein et al. [2006]), which shows results comparable to ESMs.

It is said that the history of modern EMICs began with the statistical-dynamical atmosphere model of Petoukhov ([1980]). More than ten models developed prior to Claussen et al. ([2002]) defined the term EMICs based on three factors: the number of interacting components of the climate system explicitly represented in the model, the number of processes explicitly simulated, and the detail of description. The number of models increased to 13 by the time EMICs were summarized by Claussen ([2005]). As described below, we had 15 models in the latest EMIC model intercomparison project, and this number is expected to increase further.

In this chapter, we begin with a description of EMICs and a classification and then review past model intercomparison projects to determine inter-model differences between EMICs. Finally, we summarize studies using EMICs and future possibilities, with a special focus on carbon-cycle and anthropogenic emission.

Characteristics and Classification of EMICs

EMICs simplify the atmosphere and/or ocean by reducing the number of represented processes and their dimensions. Also, because EMICs generally use coarser spatial resolution in comparison to GCMs/ESMs, they run at much faster speeds than those models. For example, as in Table 2, on a scalar-type supercomputer, the EMIC Japan Uncertainty Modelling Project Loosely Coupled Model (JUMP-LCM) (Tachiiri et al. [2010]; MIROC-lite component, Oka et al. [2011]) ran significantly faster than the Atmosphere-Ocean GCM 'MIROC4h', the high-resolution version of MIROC4. It is difficult to say how many times faster because it depends on the degree of parallelization and other factors. However, with the settings in Table 2, JUMP-LCM ran 63,000 times faster than MIROC4h, and also runs significantly faster than a GCM of a medium-resolution version of MIROC4 with the conditions shown in Table 2.

Table 2: Model specifications and benchmarks: EMIC versus GCMs (super-computer at JAMSTEC)

Model	JUMP-LCM (Tachiiri et al.[2010])	MIROC4h (Sakamoto et al.[2012])	MIROC4 (unreleased version)
Horizontal grids (atmosphere)	60 × 30	640 × 320	128 × 64
Vertical layers (atmosphere)	1	56	20
Horizontal grids (ocean)	60 × 30	1,280 × 912	256 × 224
Vertical layers (ocean)	15	48	41
Time step	36 h	Atmosphere: variable; ocean: 3 min	Atmosphere and ocean: 20 min
Components	Atmosphere and ocean: with carbon cycle, land: carbon cycle	Physics in atmosphere, ocean, land	
Computer	Scalar	Scalar	Vector
CPU used	1	20	4
Speed (s/year)	15	944,620	12,958

Hajima et al.

Hajima et al. Progress in Earth and Planetary Science 2014 1:29, doi: 10.1186/s40645-014-0029-y

Based on the method of simplification, atmospheric components of EMICs may be classified into four types: 1) statistical-dynamical models, e.g., Climate-Biosphere 3α (CLIMBER-3α) (Montoya et al.[2005]); 2) energy moisture balance models (EMBM), e.g., University of Victoria (UVic) (Weaver et al. [2001]); 3) quasi-geostrophic models, e.g., Loch–Vecode-Ecbilt-Clio-agism Model (LOVECLIM) (Driesschaert [2005]); and 4) a new fourth type using primitive equations, e.g., Fast Met Office/UK Universities Simulator (FAMOUS) (Smith [2012]). Statistical-dynamical models are based on time-averaged equations, wherein the effects of large-scale atmospheric and oceanic transient eddies are parameterized in terms of climatic means or neglected (Saltzman [1978]). The EMBMs are based on the vertically integrated energy-moisture balance equations, and the quasi-geostrophic models are based on a quasi-geostrophic approximation. For the ocean, some use frictional geostrophic models, but many adopt primitive equation models as ocean GCMs.

In the latest EMIC intercomparison project, EMIC AR5, designed for contributing to the IPCC Fifth Assessment Report (AR5) report, there were 15 participant EMICs (http://climate.uvic.ca/EMICAR5). Of these, four are statistical-dynamic, seven are EMBMs, two are quasi-geostrophic, and two are based on primitive equations. Seven models have a 3D atmosphere, up from three (of eight) in the Fourth Assessment Report (AR4) of IPCC (IPCC [2007]). For the ocean, eight use primitive equations, four use frictional-geostrophic models, and the rest are mixed-layer or box models. The model with the finest spatial resolution, UVic 2.9 (Weaver et al. [2001]), has $1.8° \times 3.6°$ grids for both the atmosphere and ocean, comparable to GCMs (the atmosphere of UVic 2.9 is, however, 2D). All but one have some type of sea ice scheme. Thirteen have carbon cycle components, 12 marine, and 10 terrestrial. Of the latter, six are dynamic vegetation models. Four models have sediment and weathering components (Eby et al. [2013]). Table 3 is a classification of these models, based on the ecosystem components.

Table 3: Classification of EMICs based on ecosystem components

		Ocean	
		No	Yes
	No	MIROC-lite (2,3)	FAMOUS (3,3)
		SPEEDO (3,3)	MESMO 1.0 (2,3)
			DCESS (box)
			GENIE (2,3)
	Yes (non-DGVM)	IAP RAS CM (3,3)	IGSM2.2 (2,1)
			JUMP-LCMa(2,3)
Land			Bern3D-LPJ (2,3)
			CLIMBER-2.4 (3,2)
			CLIMBER-3 (3,3)
	Yes (DVGM)	-	LOVECLIM 1.2 (3,3)

			UMD 2.0 (3,2)
			Uvic 2.9 (2,3)

[a]MIROC-lite-LCM in Eby et al. ([2013]). Compiled from Table one of Eby et al. ([2013]), based on the presence of ocean/land ecosystem components. Numbers in parentheses after model names indicate dimensions of atmosphere and of ocean. For references on each model, see Eby et al. ([2013]).

Hajima et al.

Hajima et al. Progress in Earth and Planetary Science 2014 1:29, doi: 10.1186/s40645-014-0029-y

Zonal mean surface air temperature in the present climate is well represented by EMICs, at least to a degree similar to GCMs (IPCC [2007]). However, it is generally difficult for EMICs to represent the spatial distribution of precipitation, and so it is not easy to couple them to vegetation models. This is particularly so for EMBMs. Figure 4 shows JUMP-LCM output and a comparison with observation for atmospheric temperature and precipitation. The spatial distribution of temperature is relatively well represented, even by the EMIC with a 2D atmosphere (Figure 4c). However, Figure 4d, e,f demonstrates the difficulty for EMICs in representing precipitation patterns. Figure 4d shows that the distribution of precipitation is too zonal, and there is too much precipitation in coastal areas and too little inland. The bias is smaller, but a non-negligible bias remains in the statistical-dynamical (e.g., Figure eight of Montoya et al. [2005]) and quasi-geostrophic (e.g., Goosse et al. [2010]) models. For models using primitive equations, Smith ([2012]) reported that their FAMOUS model had a similar but accentuated precipitation pattern relative to the mother GCM. Bias in precipitation patterns is critical with coupling to terrestrial vegetation models. To solve this problem, Tachiiri et al. ([2010]) used the spatial pattern of GCM output, extracted from the change in global mean surface air temperature calculated by an EMIC, to run a vegetation model.

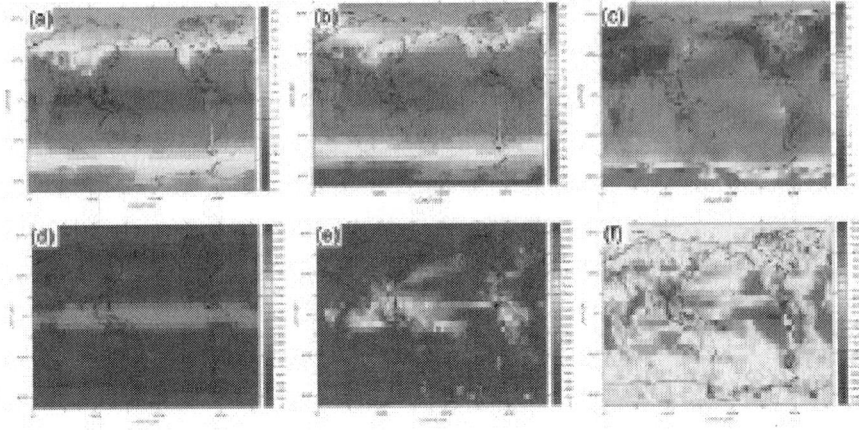

Figure 4: Modeled and observed air temperatures and precipitations for the 1990 to 1999 mean. (a) Temperature modeled by JUMP-LCM (Tachiiri et al. [2010]). (b) Observed temperature (NCEP data). (c)(a) minus (b) (unit: degree C) (d,e,f) For precipitation (unit: mm/year).

Intercomparison of EMICs

The scope of EMICs is greatly varied. It includes large-scale ocean thermohaline circulations, biogeochemical processes, planetary atmospheres, educational use, debugging, and parameter tuning, with focused time scales from decades to several million years (Claussen [2005]). Given this variety of model targets and output variables, the design of EMIC model intercomparison projects can be more difficult than that of GCMs/ESMs, in terms of what can or should be compared.

The first EMIC intercomparison project was EMIP-CO_2 (Petoukhov et al. [2005]), in which eight EMICs were involved. The modeled temperature was in relatively good agreement. However, not surprisingly, the spread of zonal precipitation was significant. In IPCC AR4, the same eight EMICs were listed (IPCC [2007]), but only half overlapped with EMIP-CO_2. Brovkin et al. ([2006]) presented their model intercomparison study results from their interesting experiments on the biogeophysical effects of historical land-cover changes during the last millennium. These results included a 0.13°C to 0.25°C global mean temperature decrease due to historical deforestation, with significant uncertainty in the zonal mean temperature and evapotranspiration.

In EMIC AR5, project protocol ranged from idealized abrupt and gradual $2xCO_2$ and $4xCO_2$, historical (last millennium), to RCP experiments. Eby et al. ([2013]) stated that similarly to ESMs, land carbon fluxes had much more variation between models than ocean carbon. Comparison of some of their results for EMICs with those for ESMs is presented in Table 1. Equilibrium climate sensitivity and its standard deviation (SD) are similar but slightly less than those of ESMs. Climate-land carbon feedback (γ_l) was similar between EMICs and ESMs, but concentration-land carbon feedback (β_l) and its SD were smaller for EMICs. In contrast, magnitudes (absolute values) of climate- and ocean concentration-carbon-cycle feedbacks (β_o and γ_o) are larger in EMICs. Most interestingly, the SDs of β_o and γ_o are more than twice those in EMICs relative to ESMs. For the last millennium simulation, there was a tendency for EMICs to underestimate the decline in surface air temperature and CO_2 between the Medieval Climate Anomaly and Little Ice Age estimated from paleoclimate reconstructions (although some ESMs used different volcanic aerosol forcing).

Regarding other studies related to EMIC AR5, Weaver et al. ([2012]) showed that EMIC representations of the strength of the present Atlantic Meridional Overturning Circulation (AMOC) are similar to those of GCMs. In addition, Zickfeld et al. ([2013]) showed the result of long-term experiments for the future to the year 3000 (forced with RCPs together with their extensions to the year 2300 and then with a fixed atmospheric CO_2 concentration and forcing from non-CO_2 greenhouse at the year 2300 levels after that), focusing on climate change commitment and reversibility. They presented the spread of temperature rise and cumulative emission for four RCPs and indicated that the meridional overturning circulation (MOC) is weakened temporarily but recovers to near-preindustrial values in most models for RCPs 2.6 to 6.0. The MOC weakening was more persistent for RCP8.5. In comparison to GCMs, the temperature increase projected by EMICs through the end of the twenty-first century was similar in low-concentration scenarios (RCPs 2.6 and 4.5) but significantly lower in RCP 8.5.

Existing and Future Studies Using EMICs

Weber ([2010]) discussed existing studies using EMICs for the transient evolution of the climate, the AMOC and hindcasting, assessment of uncertainties, and forecasting. Examples of long-duration experiments include Brovkin et al. ([2007]), Plattner et al. ([2008]), Archer et al. ([2009]), and others dealing with the role of vegetation (Tuenter et al. [2005]) and response (Claussen et al.[1999]). Studies of the long-term (e.g., up to the year 3000) commitment of the CO_2 effect (Plattner et al. [2008]; Zickfeld et al. [2013]) are included in this type.

Studies using large ensembles, many of which tuned parameters by comparison with observations, include Knutti et al. ([2002]), Forest et al. ([2002]), Hargreaves et al. ([2004]), and Annan et al. ([2005]). Annan and Hargreaves ([2010]) carried out experiments of parameter perturbation, including those related to the marine carbon cycle.

For terrestrial ecosystems, Tachiiri et al. ([2012]) used an EMIC in a parameter perturbation experiment for a vegetation model, identifying parameters that had significant impacts on land carbon uptake under global warming. In another type of carbon cycle-related study, Zickfeld et al. ([2011]) examined the nonlinearity of climate- and concentration-carbon cycle feedback using UVic-ESM.

Related to the carbon cycle, a relatively new type of study using EMICs addresses climate stabilization, through emission reduction (Matthews and Caldeira [2008]) or geoengineering (Brovkin et al. [2009]). A policy-oriented study using an EMIC and IAM was presented by Van Vuuren et al. ([2008]), and Webster et al. ([2012]) discussed climate policy targets under uncertainty using an EMIC.

An important advantage of EMICs (at least those with an EMBM-type atmosphere) is that we can easily vary climate sensitivity (Plattner et al. [2001]; Tachiiri et al. [2010]). Through this, we can assess the effect of uncertainty in climate sensitivity on ecosystems and then on the amount of emission following given concentration pathways. An example is Tachiiri et al. ([2013]) who presented the potential of constraining physical properties of Earth systems using carbon-cycle related observations (in their case, carbon emission). However, they stated that this should be done very carefully because the posterior probability distribution function of physical parameters is sensitive to the characteristics of ecosystem components.

In addition to the traditional use of EMICs (i.e., long-term integration for paleoclimatic studies or perturbed physics ensembles), EMICs are becoming important for assessing uncertainty in global climate-carbon cycle systems and for their application to treating future stabilization pathways. In the near future, given the increase in computational power, some types of EMICs may replace the climatic component of IAMs. Moreover, the increase of primitive equation models for EMIC atmospheres indicates that it may be increasingly difficult to distinguish EMICs from GCMs/ESMs by the processes included. However, this does not mean that EMIC importance is declining. On the contrary, this is the route toward a meaningful 'model hierarchy', where EMICs are effectively used for sensitivity tests and tuning parameters with large ensembles to improve the performance of the mother GCM/ESM. Due to this interactive connection between, or hybrid use of, EMICs and GCMs/ESMs, complementary relationships between these models are expected in the future.

Climate Geoengineering

Although understanding of anthropogenic climate change is steadily deepening, the political response has been slow and inadequate, which has led to a call for more drastic actions such as climate geoengineering. The newly released AR5 of IPCC ([2013]) has reviewed these schemes in a comprehensive manner for the first time in its history. In this section, we restrict the discussion to scientific aspects of geoengineering, although there are considerable controversies regarding its role in society. Interested readers are referred to the Royal Society ([2009]) and IPCC ([2012]) and references therein.

There are useful references on the science of geoengineering, such as Caldeira et al. ([2013]) and chapters 6 and 7 of IPCC Working Group 1 of AR5. In the following, we focus on the modeling of geoengineering techniques.

Definitions

According to the IPCC, geoengineering is defined as 'a broad set of methods and technologies that aim to deliberately alter the climate system in order to alleviate the impacts of climate change' (see the IPCC glossary). This is different from other responses to climate change such as mitigation (reductions of greenhouse gas emissions) or adaptation (moderating the damage by changing societal practice and behavior). Geoengineering is often divided into solar radiation management

(SRM) and carbon dioxide removal (CDR). Table 4 lists categories of proposed schemes, based on the classification of the IPCC ([2013]).

Table 4: Categories of geoengineering proposals, based on IPCC ([2013])

Category	Proposal
CDR	Afforestation and reforestation
	Bioenergy with carbon capture and storage (BECCS)
	Biochar creation and storage in soils
	Ocean fertilization by adding nutrients to surface waters
	Ocean-enhanced upwelling, bringing more nutrients to surface waters
	Land-based increased weathering
	Ocean-based increased weathering
	Direct air capture (engineering method)
SRM	Space-based methods
	Stratospheric aerosol injection
	Cloud brightening
	Surface albedo changes

Hajima *et al.*

SRM is intended to reflect some incoming solar radiation back to space, e.g., by spraying scattering aerosol particles in the stratosphere (Rasch et al. [2008]), brightening clouds (Latham [1990]), using a mirror system in space (Angel [2006]), or increasing surface albedo (Lenton and Vaughan [2009]). There are related schemes, such as a reduction in cloud forcing of cirrus clouds (Mitchell and Finnegan [2009]).

CDR (or negative emissions technology, NET) refers to a class of techniques that reduce atmospheric CO_2 concentration either by increasing natural carbon sinks or directly removing CO_2 via industrial engineering. The proposed schemes include bioenergy with carbon capture and storage (BECCS), direct air capture through chemical engineering, storing biochar in soils, and the acceleration of chemical weathering, which in nature absorbs CO_2 on a geologic time scale. Because the definition of mitigation covers the enhancement of natural sinks (e.g., afforestation and reforestation), there is some overlap between geoengineering and conventional mitigation. In fact, the simulation of RCP2.6 by an IAM, IMAGE, assumes widespread use of BECCS. Therefore, CDR may not necessarily represent an additional CO_2 reduction opportunity. Theoretically, one can achieve removal of non-CO_2 greenhouse gases, though literature on this is scarce.

Among the many proposed SRM schemes, the two most discussed are stratospheric aerosol injection and cloud brightening. The current understanding of the former is primarily based on climate response to volcanic forcing, and the latter on cloud processes and physics. For the CDR schemes, ocean iron fertilization has historically received great attention. The three schemes in this paragraph are relevant to ESS and modeling.

Characteristics

Although various techniques are grouped under the rubric of geoengineering, they have vastly different characteristics with respect to effectiveness, environmental risks, and other aspects. As Keith et al. ([2010]) summarizes, SRM has the following features. Its implementation is usually of low cost and, once initiated, can cool the climate rapidly. However, its ability to counteract climate change is imperfect and cannot offset all its impacts. Further, the science is rudimentary and all aspects of SRM are uncertain.

Take the example of stratospheric aerosol injection. This technique is believed to be capable of counteracting a doubling of CO_2 rapidly and cheaply. However, it has side effects, including ozone destruction, slowdown of the global hydrologic cycle, and reduction of electricity generation by concentrating solar power. If the injection were ever halted suddenly, it could result in a sudden rise in the global mean surface temperature by unmasking radiative forcing of greenhouse gases. This technique also fails to address ocean acidification.

The characteristics of CDR can be understood by comparison with SRM. CDR is more expensive, slower to affect the carbon cycle, but more reliable because it influences the real culprit of climate change. However, some schemes, especially those that intervene in the ecological system, are certain to have significant side effects. Cost estimates tend to be comparable to or higher than conventional mitigation.

For example, the direct air capture of CO_2 is considered to have great potential for atmospheric CO_2 reduction. Although technology exists, it is not clear whether it can be operated on an industrial scale because the cost estimate is very uncertain, but it is at least as expensive as conventional mitigation.

Modeling of CDR

Martin ([1990]) put forth a hypothesis that iron is the limiting factor in high-nutrient, low-chlorophyll (HNLC) regions such as the Southern Ocean, the equatorial Pacific, and the North Pacific. Since then, about a dozen small-scale (O (100 km²)) *in situ* experiments have been conducted (Strong et al.[2009]). Because these intervention experiments have vindicated the iron hypothesis, there was growing interest in iron fertilization of oceans for CO_2 reduction (see also 'Atmospheric deposition of bioavailable iron in marine ecosystems' section).

Several modeling studies have evaluated ocean fertilization with varying degrees of sophistication. Earlier studies, which used biogeochemical cycle models without the explicit iron cycle, implicitly modeled the effect of iron fertilization, e.g., by depleting near-surface phosphate (Sarmiento and Orr [1991]). These studies tended to report optimistic potentials for atmospheric CO_2 drawdown, and some reported approximately 100 ppm of CO_2 drawdown (Joos et al. [1991]; Cao and Caldeira[2010b]). As the models improved, however, estimates of the potential were revised downward in most studies. For example, Aumont and Bopp ([2006]) ran a biogeochemical cycle model with the iron cycle, obtaining a drawdown of 33 ppm (Aumont and Bopp [2006]). Similarly, Sarmiento et al. ([2010]) reported 42 ppm. One reason for such differences is that there are non-iron limiting factors such as silicate and light, and carbon export to the deep ocean can only occur during the growing season (Aumont and Bopp [2006]).

In addressing the amount of atmospheric removal, we must consider the 'rebound effect'. When human activities cause CO_2 emissions into the atmosphere, only about half remains there, with the rest absorbed by terrestrial and oceanic sinks. When human activity removes CO_2 from the atmosphere, the opposite occurs. To reduce the CO_2 concentration by 1 ppm, for example, one needs to take up an amount of CO_2 equivalent to approximately 2 ppm (assuming that the airborne fraction is approximately 0.5). This has been termed the rebound effect, and one must be careful about the efficacy of CDR (Cao and Caldeira [2010a]).

The ocean fertilization potentials reported above account for the rebound effect from the oceans but exclude the effect from land, thereby overestimating the potential.

Another use of ESMs is to calculate required emission reductions compatible with a certain target RCP. Cumulative emissions compatible with RCP2.6 are estimated at 272 ± 101 PgC for the period 2012 to 2100 (chapter 6 of IPCC Working Group 1 of AR5, IPCC [2013]). Some results suggest that sustained negative emissions, such as those from BECCS, are required.

Modeling of SRM

Process studies and engineering analyses have indicated that it would be feasible to conduct SRM for canceling a CO_2 doubling and that the current cost of implementation is inexpensive, at least for stratospheric aerosol injection. Significant uncertainty remains, however, as to its efficacy of counteracting climate change and its side effects. Modeling can therefore make an important contribution to the evaluation of geoengineering schemes.

The situation of SRM modeling was rudimentary until the advent of the Geoengineering Model Intercomparison Project (GeoMIP), which is related to CMIP5 activities. One of the motivations for this project was the impact of geoengineering on the Asian summer monsoon (Kravitz et al. [2011]). Some models implied a substantial decrease in monsoon precipitation, while others suggested the opposite. However, the number of models in the exercise was small and scenario specifications varied, complicating the interpretation. Obviously, a more systematic approach to modeling was needed.

The initial GeoMIP included the following four experiments:

G1: Cancel the warming from instantaneous CO_2 quadrupling with a simultaneous decrease of the solar constant

G2: Counteract warming from CO_2 increase at 1% per year by steadily reducing the solar constant

G3: Starting in 2020, offset RCP4.5 forcings by gradually increasing the amount of SO_2 or sulfate injected either at the equator or globally

G4: Starting in 2020, inject a constant amount of SO_2 at a rate of 5 Tg-SO_2 per year to partially counteract the RCP4.5 forcing.

This set of experiments had variable scenario complexities to attract many modeling groups, with two experiments focusing on the solar constant and the other two on the injection of sulfates. In CMIP5, most models used an externally specified optical depth to represent volcanic cooling, but models in GeoMIP represent a variety of approaches, from the CMIP5-type approach to directly simulating stratospheric chemistry and aerosols.

In the following, we summarize the 12-model G1 experiment by Kravitz et al. ([2013a]), which confirmed basic results of previous studies. Twelve fully coupled AOGCMs were included, 11 of which came with a land ecosystem model. The required solar constant reduction was model-dependent, between 3.5% and 5.0%.

The G1 results show that a reduction of solar insolation can largely offset the temperature changes but leave the polar region warmer (inter-model average 0.8 K) and the tropics colder (inter-model average −0.3 K) than preindustrial levels (Figure 5). This is because a reduction of solar insolation by the same fraction led to a large-magnitude reduction in equatorial regions. Similarly, net precipitation (precipitation minus evaporation) induced by quadrupling CO_2 could be mostly offset by the reduction of solar insolation, although there was less precipitation in some tropical regions. The tropical precipitation is explained by changes in moist static stability. Quadrupling CO_2 increases net primary productivity because of CO_2 fertilization effects. In the G1 results, net primary productivity increases a little more, because geoengineering creates an artificial environment in which the CO_2 level is elevated and climate change is reduced.

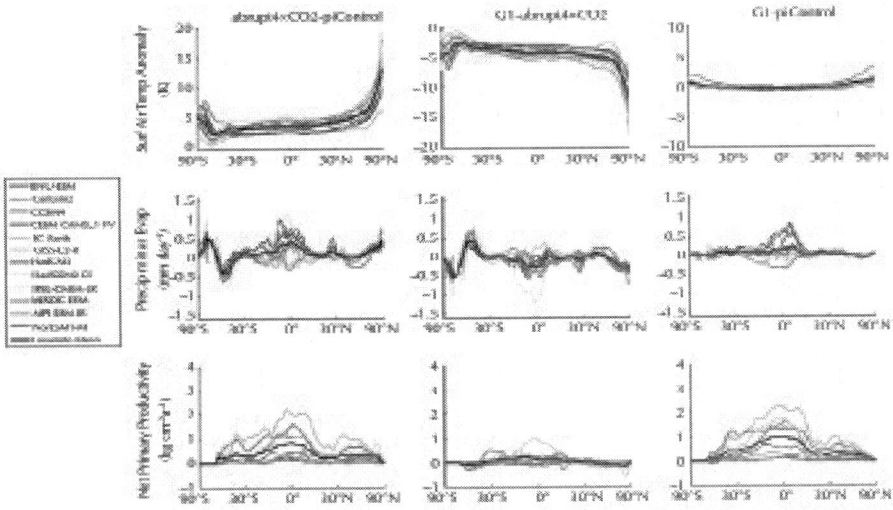

Figure 5: Impacts of solar radiation management evaluated by ESMs. Zonal-mean anomalies of surface air temperature in kelvin (land and ocean; 12 models), precipitation minus evaporation in millimeters per day (land average; 12 models), and terrestrial net primary productivity in kg C per m² per year (land average; 8 models) for all available models. All values are averages over years 11 to 50 of the simulations. Horizontal axis scales with cosine of latitude. Adapted from Figure one of Kravitz et al. ([2013a]): Climate model response from the Geoengineering Model Intercomparison Project (GeoMIP), J Geophys Res Atmos 118, pp. 8320-8332.

GeoMIP has investigated other topics related to climate geoengineering, such as changes in the hydrologic cycle, extreme events, stratospheric ozone loss, and impacts on agriculture and the cryosphere. The results have been published in a special issue of the *Journal of Geophysical Research* (Kravitz et al. [2013b]). GeoMIP research teams are discussing the next round, which includes experiments on cloud brightening.

Remaining Uncertainties

Although initial evaluations of CDR and SRM techniques are useful, there remain substantial uncertainties in various types of geoengineering method. GeoMIP made great progress in stratospheric aerosol injection and had a plan for cloud brightening as well. Nevertheless, there have been no model intercomparison projects for CDR. If governments and society were to consider such an option, a more systematic modeling exercise would be needed.

It is important to recognize that there are also substantial uncertainties in CMIP5-class climate models. Driscoll et al. ([2012]) examined the response to volcanic forcings in CMIP5 models. They identified weaker-than-observation responses of the stratospheric polar vortex and less warming of Eurasia following a volcanic eruption. This casts doubt on the dynamic responses simulated in GeoMIP.

Similarly, cloud processes are full of uncertainties, and clouds are the key factor in determining climate sensitivity. Better geoengineering simulation requires further development of ESMs.

Ocean Co$_2$ Uptake

Role of Oceans as the Largest Co$_2$ Reservoir

The ocean, like the terrestrial biosphere, is a major sink of atmospheric CO_2. Quantitative analysis shows that the ocean contains about 38,000 PgC. This is about 16 times that in the terrestrial biosphere and 60 times that in the preindustrial atmosphere, i.e., at a time before atmospheric CO_2 content was altered by the increased burning of fossil fuels from human activities (Post et al. [1990]). Therefore, the ocean is the largest carbon reservoir and is critical in determining atmospheric CO_2 concentration and thereby the global radiation balance and climate (IPCC [2007]). However, the carbon sink from oceanic mixing is slow and requires centuries to convect to the deep ocean (Broecker et al. [1982]). Hence, changes in atmospheric CO_2 concentration induced by oceanic modulations also occur on century time scales (Broecker and Peng [1986]).

Carbon Cycle in the Ocean

The carbon cycle in the ocean in terms of biogeochemical processes can be described as follows. When absorbed by seawater, most CO_2 turns into bicarbonate ions; this dissolved inorganic form stabilizes 100 times more carbon than CO_2 in molecular form. At the sea surface, marine ecosystems such as phytoplankton consume the CO_2 through photosynthesis and enhance oceanic CO_2 uptake. After the consumption of CO_2 by marine organisms, their detritus convects and stabilizes carbon in the middle and deep ocean interior. Since the marine organisms are important, this downward transport of CO_2 from the surface to the middle and deep ocean is the so-called 'CO_2 capture and storage (CCS)' and 'biological pump' (Volk and Hoffert [1985]; Longhurst and Harrison [1989]).

Takahashi et al. ([2002], [2009]) analyzed global distributions of monthly and annual net sea-air CO_2 flux based on nearly a million measurements of the sea surface partial pressure of CO_2 (pCO_2) since late the 1950s. They estimated the annual net uptake of CO_2 by the global oceans at approximately 2.2 PgC. Strong CO_2 sink areas were seen in the transition zone between the subtropical gyre and subpolar waters, around 60°S to 40°S in the Southern Ocean and 40°N to 60°N in the North Atlantic, owing to low pCO_2 waters formed by juxtaposition of the cooling of warm waters with biological drawdown of pCO_2 in the nutrient-rich subpolar waters. Nevertheless, the most intense CO_2 source areas are the eastern equatorial Pacific and northwestern Arabian Sea. The tropical Atlantic and Indian oceans and northwestern subarctic Pacific are also prominent source areas. During La Niña events, the equatorial Pacific could be a stronger CO_2 source, in response to local upwelling of CO_2-rich waters (Feely et al. [1999]).

Anthropogenic Co_2 Uptake

Sabine et al. ([2004]) estimated a global oceanic anthropogenic CO_2 sink for the period of 1800 to 1994 for 99 to 137 PgC based on the most accurate and comprehensive measurements of inorganic carbon, oxygen, nutrients, and chlorofluorocarbons. These measurements were made in the 1990s during two international ocean research programs, the World Ocean Circulation Experiment (WOCE) and Joint Global Ocean Flux Study (JGOFS). The oceanic sink accounted for approximately 48% of total fossil-fuel and cement-manufacturing emissions for the about 200 years since the beginning of the industrial period. This could potentially increase atmospheric CO_2 to approximately 55 ppm higher than the present level, if released from the ocean.

In contrast to the oceanic CO_2 sink, the terrestrial biosphere has been a net source of CO_2 to the atmosphere of about 11 to 67 PgC, because

CO_2 emissions from land-use change (100 to 180 PgC) overwhelm the CO_2 uptake of the terrestrial biosphere (61 to 141 PgC) (Sabine et al. [2004]). High anthropogenic CO_2 concentrations are found in the North Atlantic and between 50°S and 14°S, representing 23% and 40%, respectively, of the global oceanic anthropogenic CO_2. About 30% of the anthropogenic CO_2 is found at depths shallower than 200 m. Nearly 50% is at depths above 400 m and is rarely observed below 1,000 m.

The Southern Ocean

As a major sink of atmospheric CO_2, the Southern Ocean absorbs CO_2 at an annual rate of about 1.5 PgC (IPCC [2007]), although this rate has been decreasing. Recent trends in the Southern Hemisphere tropospheric circulation can be interpreted as a bias toward high-index polarity of the Southern Annular Mode (SAM), with stronger westerly flow encircling the polar cap (Thompson and Solomon [2002]). The largest and most significant tropospheric trends can be traced to recent variations in the lower stratospheric polar vortex, which are largely attributed to photochemical ozone losses. Surface cooling over most of Antarctica is also associated with shifts in the SAM contributed by the Antarctic ozone depletion and increasing greenhouse gases (Shindell and Schmidt [2004]).

Based on observed atmospheric CO_2 concentration and an inverse method, Le Quéré et al. ([2007]) estimated that the Southern Ocean sink of CO_2 south of 45°S weakened between 1981 and 2004 by 0.08 PgC per decade, relative to the trend expected from the large increase in atmospheric CO_2. They attributed this weakening to the observed increase in Southern Ocean winds resulting from human activities, which is projected to continue in the future. Consequences include a reduction in the efficiency of the Southern Ocean sink of CO_2 in the short term (approximately 25 years) and possibly greater stabilization of atmospheric CO_2 on a multi-century scale (Figure 6).

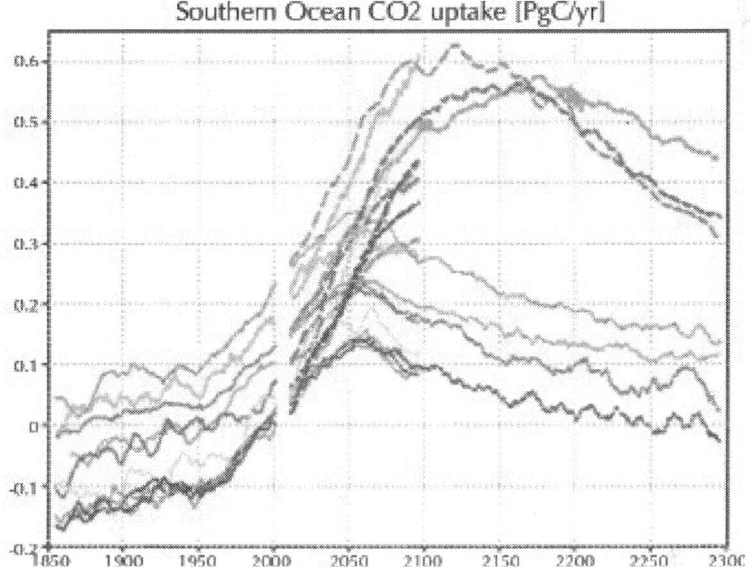

Figure 6: Time series of Southern Ocean CO2uptake (petagrams of carbon per year) simulated by CMIP5 ESMs. Monthly sea surface CO_2flux is integrated spatially to the south of 45°S and averaged temporally for 5 years before and after (10-year running mean). CO_2 concentration-driven runs of historical (thin solid lines, December 2005 and prior), RCP4.5 (thin dashed lines, January 2006 and after), and RCP8.5 (thick dashed lines, January 2006 and after). Model outputs are from CESM1-BGC (red), GFDL-ESM2G (green), GFDL-ESM2M (blue), HadGEM2-ES (cyan), IPSL-CM5A-LR (magenta), IPSL-CM5A-MR (yellow), MIROC-ESM-CHEM (orange), MIROC-ESM (purple), MPI-ESM-MR (yellow green), and bcc-csm1-1 (medium blue).

Atmospheric Deposition of Bioavailable Iron in Marine Ecosystems

Iron (Fe) is an essential micronutrient for primary production in marine ecosystems (Jickells et al.[2005]; Uematsu [2013]; see also the 'Modeling of CDR' section). Since most aquatic organisms can take up iron only in dissolved form, the amount of soluble iron is of major importance (Raiswell and Canfield [2012]). The majority of iron is delivered from arid and semiarid regions to the open ocean but is mainly in insoluble form (Mahowald et al. [2009]). Insoluble iron oxides in soils can be transformed to soluble iron in mineral aerosols via acid processing, photochemical reduction, and ligand-promoted iron dissolution during cycling of dust particles between cloud and aerosol water (Shi et al. [2012]). In addition to iron in the form of aqueous species, aerosols supply iron in colloidal or nanoparticulate forms that can be transformed into soluble iron in seawater (Baker and Croot[2010]). Compared with mineral aerosols, aerosols from combustion processes contribute to high iron solubility observed over oceans (Sholkovitz et al. [2012]). In this review, we focus on global atmospheric modeling studies of soluble iron input to the oceans. This builds on a number of earlier reviews of iron in the biogeochemical cycle (Jickells et al. [2005]; Mahowald et al. [2009]; Baker and Croot [2010]; Raiswell and Canfield [2012]; Shi et al. [2012]; Sholkovitz et al. [2012]).

Global Atmospheric Modeling Studies

The atmospheric deposition of iron in the oceans has been extensively examined in global modeling studies, and the predicted total iron deposition is in better agreement between various studies than in soluble iron deposition, owing to large uncertainties in aerosol iron solubility (i.e., the percentage of total aerosol iron that is soluble in water) (Table 5). Conventionally, global bioavailable iron deposition has been estimated with prescribed iron solubility in aerosols at 1% to 2% (Jickells et al.[2005]). To estimate the effects of the atmospheric processing of dust on iron solubility in a global transport model of mineral aerosols, Hand et al. ([2004]) used a soluble decay lifetime of 300 days to represent the conversion of insoluble iron to soluble iron at 20% iron solubility for dust transport across the Atlantic Ocean. Similarly, some global transport models use parameterizations of iron dissolution rates in mineral aerosols to fit observations of iron solubility (Luo et al. [2005]; Fan et al.[2006]; Han et al. [2012]). However, elevated iron solubility values have been measured for bulk aerosol samples when combustion aerosols are captured on the filters (Chuang et al. [2005]; Guieu et al. [2005]; Sedwick et al. [2007]). Thus, the fitted iron dissolution rate of mineral dust with observed iron solubility overestimates solubility of the mineral dust in models. Fast solution methods for atmospheric processing are useful when investigating factors affecting ecosystem and climate changes because atmospheric chemistry is the most time-consuming part of the calculations in

ESMs (Ito and Kawamiya [2010]). To increase accuracy and reduce uncertainty in the models, however, it is essential that a simplified model give results that are equivalent to more complete process-based mechanisms.

Table 5: Global iron (Fe) and soluble Fe deposition on the oceans and fractional Fe solubility

Study	Fe (Tg year-1)	Soluble Fe (Tg year-1)	Fe solubility (%)
Fan et al. ([2006])	21	2.3	11
Han et al. ([2012])	11	0.47	4.1
Ito et al. ([2012])	15	0.29	1.9
Ito ([2013])	16	0.45	2.8
Ito and Xu ([2014])	12	0.35	2.8
Jickells et al. ([2005])	16	0.16 to 0.32	1 to 2
Johnson and Meskhidze ([2013])	18a	0.26	1.4
Luo et al. ([2005])	11b	0.36 to 1.6	3.3 to 15
Luo et al. ([2008])	11b	0.21b	1.9
Luo and Gao ([2010])	11b	0.34	3.1

[a]Personal communications. [b]Mahowald et al. ([2009]).

Hajima et al.

Hajima et al. Progress in Earth and Planetary Science 2014 1:29, doi: 10.1186/s40645-014-0029-y

The chemical and physical properties of combustion and dust aerosols are different (Siefert et al. [1999]; Desboeufs et al. [2005]), and thus it is important to understand the influence of those properties on iron solubility. Luo et al. ([2008]) compiled emission factors for iron in combustion sources such as coal combustion and biomass burning. Their model results suggested that combustion sources of iron contribute a significant amount of soluble iron deposition over oceans downwind of industrialized and biomass burning regions, because of high iron solubility. It was also pointed out that biomass burning (e.g., savanna and forest fires) is an important source of soluble iron in areas with low levels of atmospheric pollutants and mineral dust (Ito [2011], [2012]). However, including combustion aerosols in global models did not show any significant improvement in comparison with observations of iron solubility (Luo et al. [2008]; Ito [2012]). By making a distinction for iron solubility between the various sources of iron in combustion aerosols, Ito ([2013]) suggested that atmospheric models improve the simulation of high iron solubility (>10%) at low iron loading (<100 ng m^{-3}) over the open ocean (Figure 7). The main mechanisms leading to enhanced iron solubility are the high iron solubility associated with oil combustion aerosols from shipping and the iron mobilization of mineral dust. The former process can produce high iron solubility at low iron loading. The latter process can transform water-insoluble iron in soils to soluble forms (e.g., ferrihydrite colloids, nanoparticles, and aqueous species) during long-range atmospheric transport. The model predicts narrower variability of iron solubility for large mass concentrations (100 to 10,000 ng m^{-3}) than observed, under a variety of conditions (Figure 7). These results suggest that achieving a more accurate simulation of iron solubility (0.1% to 10%) for a wide range of mass concentrations (10 to 10,000 ng m^{-3}) has important implications for the range of model-based estimates of soluble iron deposition and its response to environmental change.

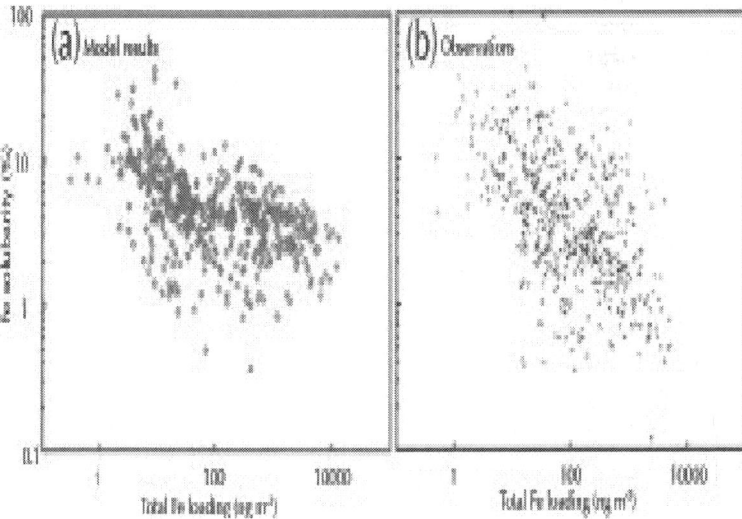

Figure 7: Atmospheric loading of total aerosol iron (ng m−3) versus percent of soluble iron in total iron. (a) Model results (red circles) of Ito ([2013]) and (b) observations (black crosses) over the oceans from 2001 to 2006. Aerosol iron solubility is plotted against iron loading on a $\log_{10}$ scale to emphasize values at low solubility in Figure eight **(a)** and **(b)** of Ito ([2013]).

Mineral dissolution rates depend on several kinetic parameters (e.g., temperature, pH, and degree of solution saturation) (Zhuang et al. [1992]). Based on laboratory experiments for iron oxides (i.e., hematite), Meskhidze et al. ([2005]) used the parameterization of proton-promoted acid mobilization (Lasaga et al. [1994]) for mineral aerosols in a Lagrangian box model, which involves a thermodynamic equilibrium module to estimate acidity in the aqueous phase of hygroscopic particles. A chemical transport model that implemented the iron dissolution scheme for dust aerosols suggested significant acid mobilization of iron in the East Asia outflow over the North Pacific, because aerosol water becomes very acidic (pH <2) due to air pollution (Solmon et al. [2009]). However, such a highly acidic condition is very rare for mineral dust in the atmosphere because alkaline gases and minerals neutralize the acidic species in most cases (Ito and Feng [2010]; Johnson et al. [2010]).

It is obvious that modeling must follow experimental and mechanistic understanding. Laboratory experiments for different dust source samples suggest much faster iron dissolution rates than those used in the explicit dissolution scheme (Mackie et al. [2005]; Shi et al. [2011a]). The iron mobilization model is based on the assumption that the mineral dissolution behaves similarly to simple iron oxides under acidic conditions. However, mineral dust is composed of variable amounts of clay minerals (e.g., illite, kaolinite, and smectite), carbonates (e.g., calcite), quartz, oxides (e.g., hematite), feldspars, and evaporite minerals (e.g., gypsum) (Claquin et al. [1999]; Nickovic et al.[2012]). Thus, aerosol mineralogy, especially the chemical form of iron in aerosols, can be a critical factor for iron dissolution (Cwiertny et al. [2008]; Journet et al. [2008]; Shi et al. [2011b]). Further, the dissolution of iron oxides depends critically on particle size (Kraemer [2004]; Rubasinghege et al.[2010]; Lanzl et al. [2012]). Based on laboratory experiments for specific mineral samples (Journet et al. [2008]; Shi et al. [2011b]) and soil and dust samples (Mackie et al. [2005]; Shi et al. [2011a]), Ito and Xu ([2014]) improved the treatment of iron in mineral dust and its dissolution scheme under acidic conditions to emulate dissolution curves for African, Asian, and Australian samples. This more complete understanding of the processes controlling iron dissolution improves the predictive capability for the wide variability of iron solubility (1% to 10%) over the oceans of the Northern Hemisphere (Ito and Xu [2014]).

Future Outlook: Interactions of Mineral Aerosols with Organics

To facilitate more accurate parameterization of iron dissolution in atmospheric water, it is desirable that laboratory conditions be as representative as possible of ambient conditions. Low molecular weight dicarboxylic acids (e.g., oxalic acids) are recognized as ubiquitous aerosol constituents in the troposphere (Kawamura [2006]; Ervens et al. [2011]). In recent years, increased attention has been given to the organic acids for promoting iron dissolution in global models (Luo et al. [2005]; Luo and Gao [2010]; Johnson and Meskhidze [2013]). To estimate the oxalate concentration in mineral dust, Luo and Gao ([2010]) used the molar ratio of oxalate to sulfate (2%) based on observations ranging from 0.46% to 5% (Yu et al. [2005]). The observed oxalate refers to the sum of oxalic acid and oxalate in aerosol particles, since ion chromatography allows quantification of oxalate independent of its chemical form. The positive relationship between oxalate and sulfate is likely due to a common dominant formation pathway in cloud droplets (Yao et al. [2002]; Yu et al. [2005]). In cloud water, water-soluble organic compounds dissolve into the aqueous phase and form oxalic acids through oxidation (Myriokefalitakis et al. [2011]; Liu et al. [2012]; Lin et al. [2014]). Since iron dissolves slowly from mineral soils in cloud water conditions (up to 0.26% of iron solubility after 60 min contact time; Paris et al. [2011]), the effect of oxalate-promoted dissolution on dissolved iron concentrations in cloud water is not large enough to explain the variability of iron solubility in atmospheric aerosols (0.1% to 10%).

Furthermore, strong complexation of iron with organic compounds (e.g., humic substances) has been observed in rainwater (e.g., Kieber et al. [2003]; Cheize et al. [2012]). Furukawa and Takahashi ([2011]) showed that most of the oxalate is present as stable complexes in aerosol particles. These results suggest that iron in cloud water and oxalate in aerosol might not be readily available for iron-oxalate complexes but present as more stable forms. Because of a lack of knowledge of specific organic compounds and their formation rates in cloud and aerosol water, it is premature to draw conclusions regarding the nature of iron-organic complexes (Deguillaume et al. [2005]; Willey et al.[2012]). Further research on laboratory and field measurements are required to investigate the major formation process of oxalic acid and its effect on iron dissolution in mineral aerosols. Atmospheric chemical transport models are useful to validate the laboratory results for their application to ambient atmospheric conditions.

CONCLUSIONS

The ESS concept has existed for a relatively long time, but its urgent need has only been met over the last decade, being partially realized by ESMs. This is perhaps why scientists remain hesitant to acknowledge this new field as a discipline. Another reason may be that while other new scientific fields have often defined themselves by restricting subjects and methods and decomposing their subjects into components at finer spatial and temporal scales, ESS and its modeling have, by nature, an inclination toward widening of subjects and integration of existing paradigms to investigate component interactions spatially and temporally.

The authors' view is that ESS should be regarded as a framework of thought rather than an established discipline. This interpretation is perhaps more readily acceptable to many scientists. However, another question arises as to whether ESS is really a 'specialty' worth spending a scientist's entire lifetime (or a considerable portion thereof). However, given the current situation in which responding to global change is a pressing mission for earth science as a whole, there is an obvious need for a 'bridge' to connect ever-deepening traditional disciplines, thereby enabling a heuristic and holistic approach to interactions among subsystems of the global environment. It would be invaluable for Earth system scientists to create and develop a common mindset that commands a bird's-eye view of relevant fields of earth science and possibly social science. The authors believe that modeling will continue to contribute to cooperative growth of diversifying disciplines and expanding ESS.

AUTHORS' CONTRIBUTIONS

MK designed the outline of this paper and wrote the 'Introduction' and 'Conclusions' sections. TH was responsible for the 'Introduction' and 'Nitrogen cycle' sections and compiled the entire document. MW, EK, and KT contributed to the 'Ocean acidification,' 'Land-use and land-cover change,' and 'Earth system models of intermediate complexity' sections, respectively. MS and SW contributed to the 'Climate geoengineering' section. HO and AI were respectively responsible for the sections 'Ocean CO_2 uptake' and 'Atmospheric deposition of bioavailable iron in marine ecosystems'. All authors read and approved the final manuscript.

ACKNOWLEDGEMENTS

This work was supported by SOSEI, the 'Program for Risk Information on Climate Change', by the Ministry of Education, Culture, Sports, Science and Technology of Japan. The Earth Simulator and JAMSTEC Super Computing System were used for outputs introduced in the article. The authors are grateful to Kohei Ishihara who helped with CMIP5 data collection and server management. The author of the

'Earth system models of intermediate complexity' section thanks Dr. A. Oka and Dr. M. Yoshimori of the Atmosphere and Ocean Research Institute (AORI) of the University of Tokyo for their helpful comments on an earlier version of the manuscript. Thanks are also due to Dr. T. T. Sakamoto (AORI) for his comments on Table 2.

REFERENCES

1.   Anav A, Friedlingstein P, Kidston M, Bopp L, Ciais P, Cox P, Jones C, Jung M, Myneni R, Zhu Z (2013) Evaluating the land and ocean components of the global carbon cycle in the CMIP5 Earth system models. J Climate 26(18):6801-6843

2.   Andrews T, Gregory JM, Webb MJ, Taylor KE (2012) Forcing, feedbacks and climate sensitivity in CMIP5 coupled atmosphere-ocean climate models. Geophys Res Lett 39(9):L09712 doi:10.1029/2012GL051607

3.   Angel R (2006) Feasibility of cooling the Earth with a cloud of small spacecraft near the inner Lagrange point (L1). P Natl Acad Sci 103(46):17184-17189

4.   Annan J, Hargreaves J (2010) Efficient identification of ocean thermodynamics in a physical/biogeochemical ocean model with an iterative importance sampling method. Ocean Model 32(3):205-215

5.   Annan J, Hargreaves J, Edwards N, Marsh R (2005) Parameter estimation in an intermediate complexity Earth system model using an ensemble Kalman filter. Ocean Model 8(1):135-154

6.   Archer D, Eby M, Brovkin V, Ridgwell A, Cao L, Mikolajewicz U, Caldeira K, Matsumoto K, Munhoven G, Montenegro A (2009) Atmospheric lifetime of fossil fuel carbon dioxide. Annu Rev Earth Pl Sc 37(1):117-134

7.   Arora VK, Boer GJ, Friedlingstein P, Eby M, Jones CD, Christian JR, Bonan G, Bopp L, Brovkin V, Cadule P, Hajima T, Ilyina T, Lindsay K, Tjiputra JF, Wu T (2013) Carbon-concentration and carbon-climate feedbacks in CMIP5 Earth system models. J Climate 26(15):5289-5314 doi:10.1175/jcli-d-12-00494.1

8.   Artioli Y, Blackford JC, Butenschon M, Holt JT, Wakelin SL, Thomas H, Borges AV, Allen JI (2012) The carbonate system in

the North Sea: sensitivity and model validation. J Marine Syst 102:1-13 doi:10.1016/J.Jmarsys.2012.04.006

9. Asrar G, Busalacchi A, Hurrell J (2012) Developing plans and priorities for climate science in service to society. Eos Trans AGU 93(12):128

10. Aumont O, Bopp L (2006) Globalizing results from ocean in situ iron fertilization studies. Global Biogeochem Cy 20(2):GB2017 doi:10.1029/2005GB002591

11. Baker A, Croot P (2010) Atmospheric and marine controls on aerosol iron solubility in seawater. Mar Chem 120(1):4-13

12. Bala G, Caldeira K, Wickett M, Phillips T, Lobell D, Delire C, Mirin A (2007) Combined climate and carbon-cycle effects of large-scale deforestation. P Natl Acad Sci 104(16):6550-6555

13. Bates NR (2007) Interannual variability of the oceanic CO_2 sink in the subtropical gyre of the North Atlantic Ocean over the last 2 decades. J Geophys Res-Oceans 112(C9):Artn C09013 doi:10.1029/2006jc003759

14. Bates NR, Peters AJ (2007) The contribution of atmospheric acid deposition to ocean acidification in the subtropical North Atlantic Ocean. Mar Chem 107(4):547-558 doi:10.1016/J. Marchem.2007.08.002

15. Bodirsky B, Popp A, Weindl I, Dietrich J, Rolinski S, Scheiffele L, Schmitz C, Lotze-Campen H (2012) N_2O emissions from the global agricultural nitrogen cycle-current state and future scenarios. Biogeosciences 9(10):4169-4197

16. Bonan GB (2008) Forests and climate change: forcings, feedbacks, and the climate benefits of forests. Science 320(5882):1444-1449

17. Bonan GB, Levis S (2010) Quantifying carbon-nitrogen feedbacks in the Community Land Model (CLM4). Geophys Res Lett 37:L07401 doi:10.1029/2010gl042430

18. Bouwman A, Van Drecht G, Van der Hoek K (2005) Global and regional surface nitrogen balances in intensive agricultural production systems for the period 1 ¼—2¼3¼. Pedosphere 15(2):137-155

19. Boysen LR, Brovkin V, Arora VK, Cadule P, de Noblet-Ducoudré N, Kato E, Pongratz J, Gayler V (2014) Global and regional effects of land-use change on climate in 21st century simulations with interactive carbon cycle. Earth Sys Dynam Discuss 5:443-472

20. Broecker WS, Peng TH, Beng Z (1982) Tracers in the sea. Lamont-Doherty Geological Observatory, Columbia University, New York.

21. Broecker WS, Peng TH (1986) Carbon cycle; 1985 glacial to interglacial changes in the operation of the global carbon cycle. Radiocarbon 28(2A):309-327

22. Broecker W, Clark E (2001) A dramatic Atlantic dissolution event at the onset of the last glaciation. Geochem Geophy Geosy 2(11):2001GC000185 doi:10.1029/2001GC000185

23. Brovkin V, Boysen L, Arora V, Boisier J, Cadule P, Chini L, Claussen M, Friedlingstein P, Gayler V, Van Den Hurk B (2013) Effect of anthropogenic land-use and land-cover changes on climate and land carbon storage in CMIP5 projections for the twenty-first century. J Climate 26(18):6859-6881

24. Brovkin V, Claussen M, Driesschaert E, Fichefet T, Kicklighter D, Loutre M-F, Matthews H, Ramankutty N, Schaeffer M, Sokolov A (2006) Biogeophysical effects of historical land cover changes simulated by six Earth system models of intermediate complexity. Clim Dynam 26(6):587-600

25. Brovkin V, Ganopolski A, Archer D, Rahmstorf S (2007) Lowering of glacial atmospheric CO_2 in response to changes in oceanic circulation and marine biogeochemistry. Paleoceanography 22(4):PA4202

26. Brovkin V, Petoukhov V, Claussen M, Bauer E, Archer D, Jaeger C (2009) Geoengineering climate by stratospheric sulfur injections: Earth system vulnerability to technological failure. Climatic Change 92(3–4):243-259

27. Bruinsma J (2009) The resource outlook to 2050: by how much do land, water, and crop yields need to increase by 2050? FAO Expert meeting on 'How to feed the world in 2050': 24–26 June 2009. FAO, Rome.

28. Budyko MI (1969) The effect of solar radiation variations on the climate of the earth. Tellus 21(5):611-619

29. Cai WJ, Hu XP, Huang WJ, Murrell MC, Lehrter JC, Lohrenz SE, Chou WC, Zhai WD, Hollibaugh JT, Wang YC, Zhao PS, Guo XH, Gundersen K, Dai MH, Gong GC (2011) Acidification of subsurface coastal waters enhanced by eutrophication. Nat Geosci 4(11):766-770 doi:10.1038/Ngeo1297

30. Caldeira K, Bala G, Cao L (2013) The science of geoengineering. Annu Rev Earth Pl Sc 41:231-256

31. Caldeira K, Wickett ME (2003) Oceanography: anthropogenic carbon and ocean pH. Nature 425(6956):36 doi:10.1038/425365a

32. Caldeira K, Wickett ME (2005) Ocean model predictions of chemistry changes from carbon dioxide emissions to the atmosphere and ocean. J Geophys Res-Oceans 110(C9), doi:10.1029/2004jc002671

33. Canfield DE, Glazer AN, Falkowski PG (2010) The evolution and future of Earth's nitrogen cycle. Science 330(6001):192-196 doi:10.1126/science.1186120

34. Cao L, Caldeira K (2010) Atmospheric carbon dioxide removal: long-term consequences and commitment. Environ Res Lett 5(2):024011

35. Cao L, Caldeira K (2010) Can ocean iron fertilization mitigate ocean acidification? Climatic Change 99(1–2):303-311

36. Cheize M, Sarthou G, Croot PL, Bucciarelli E, Baudoux AC, Baker AR (2012) Iron organic speciation determination in rainwater using cathodic stripping voltammetry. Analytica Chimica Acta 736:45-54

37. Chuang P, Duvall R, Shafer M, Schauer J (2005) The origin of water soluble particulate iron in the Asian atmospheric outflow. Geophys Res Lett 32(7):L07813

38. Claquin T, Schulz M, Balkanski Y (1999) Modeling the mineralogy of atmospheric dust sources. J Geophys Res 104(D18):22243-22256

39. Claussen M (2005) Table of EMICs: Earth system models of intermediate complexity. Institut für Klimafolgenforschung, Potsdam.

40. Claussen M, Brovkin V, Ganopolski A (2001) Biogeophysical versus biogeochemical feedbacks of large-scale land cover change. Geophys Res Lett 28(6):1011-1014

41. Claussen M, Kubatzki C, Brovkin V, Ganopolski A, Hoelzmann P, Pachur HJ (1999) Simulation of an abrupt change in Saharan vegetation in the Mid-Holocene. Geophys Res Lett 26(14):2037-2040

42. Claussen M, Mysak LA, Weaver AJ, Crucifix M, Fichefet T, Loutre MF, Weber SL, Alcamo J, Alexeev VA, Berger A, Calov R, Ganopolski A, Goosse H, Lohmann G, Lunkeit F, Mokhov II, Petoukhov V, Stone P, Wang Z (2002) Earth system models of intermediate complexity: closing the gap in the spectrum of climate system models. Clim Dynam 18(7):579-586 doi:10.1007/S00382-001-0200-1

43. Cox PM, Betts RA, Jones CD, Spall SA, Totterdell IJ (2000) Acceleration of global warming due to carbon-cycle feedbacks in a coupled climate model. Nature 408(6809):184-187 doi:10.1038/35041539

44. Cwiertny DM, Young MA, Grassian VH (2008) Chemistry and photochemistry of mineral dust aerosol. Annu Rev Phys Chem 59:27-51

45. Davin EL, de Noblet-Ducoudré N (2010) Climatic impact of global-scale deforestation: radiative versus nonradiative processes. J Climate 23(1):97-112

46. Deguillaume L, Leriche M, Desboeufs K, Mailhot G, George C, Chaumerliac N (2005) Transition metals in atmospheric liquid phases: sources, reactivity, and sensitivity parameters. Chem Rev 105:3388-3431

47. Desboeufs K, Sofikitis A, Losno R, Colin J, Ausset P (2005) Dissolution and solubility of trace metals from natural and anthropogenic aerosol particulate matter. Chemosphere 58(2):195-203

48. Doney SC, Fabry VJ, Feely RA, Kleypas JA (2009) Ocean acidification: the other CO_2 problem. Ann Rev Mar Sci 1:169-192 doi:10.1146/annurev.marine.010908.163834

49. Dore JE, Lukas R, Sadler DW, Church MJ, Karl DM (2009) Physical and biogeochemical modulation of ocean acidification in the central North Pacific. Proc Natl Acad Sci USA 106(30):12235-12240 doi:10.1073/pnas.0906044106

50. Driesschaert E (2005) Climate change over the next millennia using LOVECLIM, a new Earth system model including the polar ice sheets. In: PhD Thesis. Université Catholique de Louvain, Louvain-la-Neuve. p 214 [http://dial.academielouvain.be/] [http://dial.academielouvain.be/]

51. Driscoll S, Bozzo A, Gray LJ, Robock A, Stenchikov G (2012) Coupled Model Intercomparison Project 5 (CMIP5) simulations of climate following volcanic eruptions. J Geophys Res-Atmospheres 117:D17105 doi:10.1029/2012jd017607

52. Eby M, Weaver AJ, Alexander K, Zickfeld K, Abe-Ouchi A, Cimatoribus AA, Crespin E, Drijfhout SS, Edwards NR, Eliseev AV, Feulner G, Fichefet T, Forest CE, Goosse H, Holden PB, Joos F, Kawamiya M, Kicklighter D, Kienert H, Matsumoto K, Mokhov II, Monier E, Olsen SM, Pedersen JOP, Perrette M, Philippon-Berthier G, Ridgwell A, Schlosser A, Schneider von Deimling T, Shaffer G, Smith RS, Spahni R, Sokolov AP, Steinacher M, Tachiiri K, Tokos K, Yoshimori M, Zeng N, Zhao F (2013) Historical and idealized climate model experiments: an EMIC intercomparison. Clim Past 9:1111-1140 doi:10.5194/cp-9-1111-2013

53. Ervens B, Turpin B, Weber R (2011) Secondary organic aerosol formation in cloud droplets and aqueous particles (aqSOA): a review of laboratory, field and model studies. Atmos Chem Phys 11(21):11069-11102

54. Fan SM, Moxim WJ, Levy H (2006) Aeolian input of bioavailable iron to the ocean. Geophys Res Lett 33(7):L07602 doi:10.1029/2005GL024852

55. Feely RA, Wanninkhof R, Takahashi T, Tans P (1999) Influence of El Niño on the equatorial Pacific contribution to atmospheric CO_2 accumulation. Nature 398(6728):597-601

56. Feely RA, Doney SC, Cooley SR (2009) Ocean acidification: present conditions and future changes in a high-CO_2 world. Oceanography 22(4):36-47 doi:10.5670/Oceanog.2009.95

57. Feely RA, Sabine CL, Hernandez-Ayon JM, Ianson D, Hales B (2008) Evidence for upwelling of corrosive "acidified" water onto the continental shelf. Science 320(5882):1490-1492 doi:10.1126/Science.1155676

58. Feely RA, Sabine CL, Lee K, Berelson W, Kleypas J, Fabry VJ, Millero FJ (2004) Impact of anthropogenic CO_2 on the $CaCO_3$ system in the oceans. Science 305(5682):362-366 doi:10.1126/science.1097329

59. Findlay HS, Tyrrell T, Bellerby RGJ, Merico A, Skjelvan I (2008) Carbon and nutrient mixed layer dynamics in the Norwegian Sea. Biogeosciences 5(5):1395-1410 doi:10.5194/Bg-5-1395-2008

60. Forest CE, Stone PH, Sokolov AP, Allen MR, Webster MD (2002) Quantifying uncertainties in climate system properties with the use of recent climate observations. Science 295(5552):113-117

61. Friedlingstein P, Cox P, Betts R, Bopp L, Von Bloh W, Brovkin V, Cadule P, Doney S, Eby M, Fung I, Bala G, John J, Jones C, Joos F, Kato T, Kawamiya M, Knorr W, Lindsay K, Matthews H, Raddatz T, Rayner P, Reick C, Roeckner E, Schnitzler K, Schnur R, Strassmann K, Weaver A, Yoshikawa C, Zeng N (2006) Climate-carbon cycle feedback analysis: results from the C(4)MIP model intercomparison. J Climate 19(14):3337-3353 doi:10.1175/jcli3800.1

62. Friedlingstein P, Meinshausen M, Arora VK, Jones CD, Anav A, Liddicoat SK, Knutti R (2014) Uncertainties in CMIP5 climate projections due to carbon cycle feedbacks. J Climate 27(2):511-526

63. Friedlingstein P, Dufresne J, Cox P, Rayner P (2003) How positive is the feedback between climate change and the carbon cycle? Tellus B 55(2):692-700 doi:10.1034/j.1600-0889.2003.01461.x

64. Furukawa T, Takahashi Y (2011) Oxalate metal complexes in aerosol particles: implications for the hygroscopicity of oxalate-containing particles. Atmos Chem Phys 11:4289-4301 doi:10.5194/acp-11-4289-2011

65. Galloway JN, Dentener FJ, Capone DG, Boyer EW, Howarth RW, Seitzinger SP, Asner GP, Cleveland CC, Green PA, Holland EA, Karl DM, Michaels AF, Porter JH, Townsend AR, Vöosmarty CJ (2004) Nitrogen cycles: past, present, and future. Biogeochemistry 70(2):153-226

66. Goosse H, Brovkin V, Fichefet T, Haarsma R, Huybrechts P, Jongma J, Mouchet A, Selten F, Barriat P-Y, Campin J-M (2010) Description of the Earth system model of intermediate complexity LOVECLIM version 1.2. Geosci Model Dev Discuss 3(1):309-390

67. Gregory JM, Jones CD, Cadule P, Friedlingstein P (2009) Quantifying carbon cycle feedbacks. J Climate 22(19):5232-5250 doi:10.1175/2009jcli2949.1

68. Gruber N (1998) Anthropogenic CO_2 in the Atlantic Ocean. Global Biogeochem Cy 12(1):165-191 doi:10.1029/97gb03658

69. Gruber N, Galloway JN (2008) An Earth-system perspective of the global nitrogen cycle. Nature 451(7176):293-296 doi:10.1038/nature06592

70. Guieu C, Bonnet S, Wagener T, Loÿe-Pilot MD (2005) Biomass burning as a source of dissolved iron to the open ocean? Geophys Res Lett 32(19):L19608 doi:10.1029/2005GL022962

71. Hajima T, Tachiiri K, Ito A, Kawamiya M (2014) Uncertainty of concentration-terrestrial carbon feedback in Earth System Models. J Climate 27:3425-3445

72. Han Q, Zender CS, Moore JK, Buck CS, Chen Y, Johansen A, Measures CI (2012) Global estimates of mineral dust aerosol iron and aluminum solubility that account for particle size using diffusion-controlled and surface-area-controlled approximations. Global Biogeochem Cy 26(2):GB2038 doi:10.1029/2011GB004186

73. Hand J, Mahowald N, Chen Y, Siefert R, Luo C, Subramaniam A, Fung I (2004) Estimates of atmospheric-processed soluble iron from observations and a global mineral aerosol model: Biogeochemical implications. J Geophys Res-Atmospheres (1984–2012) 109(D17):D17205 doi:10.1029/2004JD004574

74. Hargreaves J, Annan J, Edwards N, Marsh R (2004) An efficient climate forecasting method using an intermediate complexity Earth system model and the ensemble Kalman filter. Clim Dynam 23(7–8):745-760

75. Hargreaves JC, Annan JD, Yoshimori M, Abe-Ouchi A (2012) Can the Last Glacial Maximum constrain climate sensitivity? Geophys Res Lett 39:L24702

76. Houghton R, House J, Pongratz J, van der Werf G, DeFries R, Hansen M, Le Quéré C, Ramankutty N (2012) Carbon emissions from land use and land-cover change. Biogeosciences 9(12):5125-5142 doi:10.5194/bg-9-5125-201

77. Hurtt G, Chini LP, Frolking S, Betts R, Feddema J, Fischer G, Fisk J, Hibbard K, Houghton R, Janetos A (2011) Harmonization of land-use scenarios for the period 1500–2100: 600 years of global gridded annual land-use transitions, wood harvest, and resulting secondary lands. Climatic Change 109(1–2):117-161

78. Solomon S, Qin D, Manning M, Chen Z, Marquis M, Averyt KB, Tignor M, Miller HL (eds) (2007) Climate change 2007: the

Physical Science basis. In: Contribution of Working Group I to the Fourth Assessment Report of the Intergovernmental Panel on Climate Change Cambridge University Press, Cambridge/New York. p 996

79. IPCC (2012) Meeting report of the Intergovernmental Panel on climate change, Expert Meeting on Geoengineering (IPCC Working Group III Technical Support Unit). In: Edenhofer OP-MR, Sokona Y, Field C, Barros V, Stocker TF, Dahe Q, Minx J, Mach K, Plattner G-K, Schlömer S, Hansen G, Mastrandrea M (eds). Potsdam Institute for Climate Impact Research, Potsdam, p 99

80. IPCC (2013) Climate change 2013: the Physical Science basis (contribution of Working Group I to the Fifth Assessment Report of the Intergovernmental Panel on Climate Change). In: Stocker TF, Qin D, Plattner G-K, Tignor M, Allen SK, Boschung J, Nauels A, Xia Y, Bex V, Midgley PM (eds). Cambridge University Press, Cambridge/New York, p 1535

81. Ito A, Xu L (2014) Response of acid mobilization of iron-containing mineral dust to improvement of air quality projected in the future. Atmos Chem Phys 14:3441-3459 doi:10.5194/ acp-14-3441-2014

82. Ito A (2011) Mega fire emissions in Siberia: potential supply of bioavailable iron from forests to the ocean. Biogeosciences 8(6):1679-1697

83. Ito A (2012) Contrasting the effect of iron mobilization on soluble iron deposition to the ocean in the northern and southern hemispheres. J Meteorol Soc Jpn 90A:167-188 doi:10.2151/Jmsj. 2012-A09

84. Ito A (2013) Global modeling study of potentially bioavailable iron input from shipboard aerosol sources to the ocean. Global Biogeochem Cy 27(1):1-10

85. Ito A, Feng Y (2010) Role of dust alkalinity in acid mobilization of iron. Atmos Chem Phys 10(19):9237-9250

86. Ito A, Kawamiya M (2010) Potential impact of ocean ecosystem changes due to global warming on marine organic carbon aerosols. Global Biogeochem Cy 24(1):GB1012 doi:10.1029/2009GB003559

87. Ito A, Kok JF, Feng Y, Penner JE (2012) Does a theoretical estimation of the dust size distribution at emission suggest more bioavailable iron deposition? Geophys Res Lett 39:L05807 doi:10.1029/2011gl050455

88. Jain AK, Meiyappan P, Song Y, House JI (2013) CO_2 emissions from land-use change affected more by nitrogen cycle, than by the choice of land-cover data. Glob Change Biol 19(9):2893-2906

89. Jain AK, Yang X (2005) Modeling the effects of two different land cover change data sets on the carbon stocks of plants and soils in concert with CO_2 and climate change. Global Biogeochem Cy 19(2):GB2015 doi:10.1029/2004GB002349

90. Jickells T, An Z, Andersen KK, Baker A, Bergametti G, Brooks N, Cao J, Boyd P, Duce R, Hunter K (2005) Global iron connections between desert dust, ocean biogeochemistry, and climate. Science 308(5718):67-71

91. Johnson M, Meskhidze N (2013) Atmospheric dissolved iron deposition to the global oceans: effects of oxalate-promoted Fe dissolution, photochemical redox cycling, and dust mineralogy. Geosci Model Dev 6:1137-1155

92. Johnson MS, Meskhidze N, Solmon F, Gassó S, Chuang PY, Gaiero DM, Yantosca RM, Wu S, Wang Y, Carouge C (2010) Modeling dust and soluble iron deposition to the South Atlantic Ocean. J Geophys Res-Atmospheres (1984–2012) 115(D15):D15202 doi:10.1029/2009JD013311

93. Jones C, Robertson E, Arora V, Friedlingstein P, Shevliakova E, Bopp L, Brovkin V, Hajima T, Kato E, Kawamiya M, Liddicoat S, Lindsay K, Reick CH, Roelandt C, Segschneider J, Tjiputra J (2013) Twenty-first-century compatible CO_2 emissions and airborne fraction simulated by CMIP5 Earth system models under four representative concentration pathways. J Climate 26(13):4398-4413 doi:10.1175/jcli-d-12-00554.1

94. Joos F, Sarmiento JL, Siegenthaler U (1991) Estimates of the effect of Southern Ocean iron fertilization on atmospheric CO_2 concentrations. Nature 349:772-775

95. Journet E, Desboeufs KV, Caquineau S, Colin JL (2008) Mineralogy as a critical factor of dust iron solubility. Geophys Res Lett 35(7):L07805 doi:10.1029/2007GL031589

96. Kawamura K (2006) Composition and transformation of organic aerosols in the atmosphere. Chikyukagaku (Geochemistry) 40:65-82 (in Japanese)

97. Keith DW, Parson E, Morgan MG (2010) Research on global sun block needed now. Nature 463(7280):426-427

98. Kerr RA (2011) Time to adapt to a warming world, but where's the science? Science 334:1052-1053

99. Kieber R, Hardison DR, Whitehead RF, Willey JD (2003) Photochemical production of Fe(II) in rainwater. Envir Sci Tech 37:4610-4616

100. Klein Goldewijk K, Beusen A, Janssen P (2010) Long-term dynamic modeling of global population and built-up area in a spatially explicit way: HYDE 3.1. The Holocene 20(4):565-573

101. Klein Goldewijk K, Beusen A, Van Drecht G, De Vos M (2011) The HYDE 3.1 spatially explicit database of human-induced global land-use change over the past 12,000 years. Global Ecol Biogeogr 20(1):73-86

102. Klein Goldewijk K, Verburg PH (2013) Uncertainties in global-scale reconstructions of historical land use: an illustration using the HYDE data set. Landscape ecology 28(5):861-877

103. Knutti R, Stocker TF, Joos F, Plattner G-K (2002) Constraints on radiative forcing and future climate change from observations and climate model ensembles. Nature 416(6882):719-723

104. Knutti R, Masson D, Masson A (2013) Climate model genealogy: generation CMIP5 and how we got there. Geophys Res Lett 40:1194-1199 doi:10.1002/grl.50256

105. Kraemer SM (2004) Iron oxide dissolution and solubility in the presence of siderophores. Aquat Sci 66(1):3-18

106. Kravitz B, Caldeira K, Boucher O, Robock A, Rasch PJ, Alterskjær K, Karam DB, Cole JN, Curry CL, Haywood JM (2013) Climate model response from the Geoengineering Model Intercomparison Project (GeoMIP). J Geophys Res-Atmospheres 118(15):8320-8332

107. Kravitz B, Robock A, Forster PM, Haywood JM, Lawrence MG, Schmidt H (2013) An overview of the Geoengineering Model Intercomparison Project (GeoMIP). J Geophys Res-Atmospheres 118(23):13103-13107

108. Kravitz B, Robock A, Boucher O, Schmidt H, Taylor KE, Stenchikov G, Schulz M (2011) The geoengineering model intercomparison project (GeoMIP). Atmos Sci Lett 12(2):162-167

109. Lanzl CA, Baltrusaitis J, Cwiertny DM (2012) Dissolution of hematite nanoparticle aggregates: influence of primary particle size, dissolution mechanism, and solution pH. Langmuir 28(45):15797-15808

110. Lasaga AC, Soler JM, Ganor J, Burch TE, Nagy KL (1994) Chemical-weathering rate laws and global geochemical cycles. Geochim Cosmochim Acta 58(10):2361-2386

111. Latham J (1990) Control of Global Warming. Nature 347(6291):339-340 doi:10.1038/347339b0

112. Le Quéré C, Rödenbeck C, Buitenhuis ET, Conway TJ, Langenfelds R, Gomez A, Labuschagne C, Ramonet M, Nakazawa T, Metzl N (2007) Saturation of the Southern Ocean CO_2 sink due to recent climate change. Science 316(5832):1735-1738

113. Le Quéré C, Andres RJ, Boden T, et al. (2013) The global carbon budget 1959-2011. Earth Sys Sci Data 5:165-185 doi:10.5194/essd-5-165-2013

114. Lenton TM, Vaughan NE (2009) The radiative forcing potential of different climate geoengineering options. Atmos Chem Phys 9(15):5539-5561

115. Lin G, Sillman S, Penner J, Ito A (2014) Global modeling of SOA: the use of different mechanisms for aqueous phase formation. Atmos Chem Phys 14:5451-5475

116. Liu JF, Horowitz LW, Fan SM, Carlton AG, Levy H (2012) Global in-cloud production of secondary organic aerosols: Implementation of a detailed chemical mechanism in the GFDL atmospheric model AM3. J Geophys Res-Atmospheres 117:D15303 doi:10.1029/2012jd017838

117. Lobell D, Bala G, Duffy P (2006) Biogeophysical impacts of cropland management changes on climate. Geophys Res Lett 33:L06708 doi:10.1029/2005GL025492

118. Longhurst AR, Glen Harrison W (1989) The biological pump: profiles of plankton production and consumption in the upper ocean. Prog Oceanogr 22(1):47-123

119. Luo C, Gao Y (2010) Aeolian iron mobilisation by dust–acid interactions and their implications for soluble iron deposition to the ocean: a test involving potential anthropogenic organic acidic species. Environ Chem 7(2):153-161

120. Luo C, Mahowald N, Bond T, Chuang PY, Artaxo P, Siefert R, Chen Y, Schauer J (2008) Combustion iron distribution and deposition. Global Biogeochem Cy 22:GB1012 doi:10.1029/2007GB002964

121. Luo C, Mahowald N, Meskhidze N, Chen Y, Siefert R, Baker A, Johansen A (2005) Estimation of iron solubility from observations and a global aerosol model. J Geophys Res 110:D23307 doi:10.1029/2005JD006059

122. Mackie DS, Boyd PW, Hunter KA, McTainsh GH (2005) Simulating the cloud processing of iron in Australian dust: pH and dust concentration. Geophys Res Lett 32:L06809 doi:10.1029/2004GL022122

123. Mahowald NM, Engelstaedter S, Luo C, Sealy A, Artaxo P, Benitez-Nelson C, Bonnet S, Chen Y, Chuang PY, Cohen DD (2009) Atmospheric iron deposition: global distribution, variability, and human perturbations. Ann Rev Mar Sci 1:245-278

124. Martin JH (1990) Glacial-interglacial CO_2 change: the iron hypothesis. Paleoceanography 5(1):1-13

125. Matthews HD, Caldeira K (2008) Stabilizing climate requires near-zero emissions. Geophys Res Lett 35(4):L04705 doi:10.1029/2007GL032388

126. McGuire AD, Sitch S, Clein JS, Dargaville R, Esser G, Foley J, Heimann M, Joos F, Kaplan J, Kicklighter DW, Meier RA, Melillo JM, Moore B, Prentice IC, Ramankutty N, Reichenau T, Schloss A, Tian H, Williams LJ, Wittenberg U (2001) Carbon balance of the terrestrial biosphere in the twentieth century: analyses of CO_2, climate and land use effects with four process-based ecosystem models. Global Biogeochem Cy 15(1):183-206 doi:10.1029/2000gb001298

127. McNeil BI, Matear RJ (2008) Southern Ocean acidification: a tipping point at 450-ppm atmospheric CO_2. P Natl Acad Sci 105(48):18860-18864

128. Meiyappan P, Jain AK (2012) Three distinct global estimates of historical land-cover change and land-use conversions for over 200 years. Frontiers of Earth Science 6(2):122-139

129. Menon S, Denman KL, Brasseur G, Chidthaisong A, Ciais P, Cox PM, Dickinson RE, Hauglustaine D, Heinze C, Holland E (2007) Couplings between changes in the climate system and biogeochemistry. Ernest Orlando Lawrence Berkeley National Laboratory, Berkeley.

130. Merico A, Tyrrell T, Cokacar T (2006) Is there any relationship between phytoplankton seasonal dynamics and the carbonate system? J Marine Syst 59(1–2):120-142

131. Meskhidze N, Chameides W, Nenes A (2005) Dust and pollution: a recipe for enhanced ocean fertilization? J Geophys Res 110:D03301 doi:10.1029/2004JD005082

132. Mitchell DL, Finnegan W (2009) Modification of cirrus clouds to reduce global warming. Environ Res Lett 4(4):045102

133. Montoya M, Griesel A, Levermann A, Mignot J, Hofmann M, Ganopolski A, Rahmstorf S (2005) The earth system model of intermediate complexity CLIMBER-3α. Part I: description and performance for present-day conditions. Clim Dynam 25(2–3):237-263

134. Myriokefalitakis S, Tsigaridis K, Mihalopoulos N, Sciare J, Nenes A, Kawamura K, Segers A, Kanakidou M (2011) In-cloud oxalate formation in the global troposphere: a 3-D modeling study. Atmos Chem Phys 11(12):5761-5782

135. (1988) Earth system science: a closer view. National Aeronautics and Space Administration, Washington, D.C..

136. Nickovic S, Vukovic A, Vujadinovic M, Djurdjevic V, Pejanovic G (2012) Technical note: high-resolution mineralogical database of dust-productive soils for atmospheric dust modeling. Atmos Chem Phys 12(2):845-855

137. Oka A, Tajika E, Abe-Ouchi A, Kubota K (2011) Role of the ocean in controlling atmospheric CO_2 concentration in the course of global glaciations. Clim Dynam 37(9–10):1755-1770

138. Orr JC, Fabry VJ, Aumont O, Bopp L, Doney SC, Feely RA, Gnanadesikan A, Gruber N, Ishida A, Joos F, Key RM, Lindsay K, Maier-Reimer E, Matear R, Monfray P, Mouchet A, Najjar RG, Plattner GK, Rodgers KB, Sabine CL, Sarmiento JL, Schlitzer R, Slater RD, Totterdell IJ, Weirig MF, Yamanaka Y, Yool A (2005) Anthropogenic ocean acidification over the twenty-first century

and its ximpact on calcifying organisms. Nature 437(7059):681-686 doi:10.1038/nature04095

139. Paris R, Desboeufs K, Journet E (2011) Variability of dust iron solubility in atmospheric waters: investigation of the role of oxalate organic complexation. Atmos Environ 45(36):6510-6517

140. Petoukhov V, Claussen M, Berger A, Crucifix M, Eby M, Eliseev A, Fichefet T, Ganopolski A, Goosse H, Kamenkovich I (2005) EMIC Intercomparison Project (EMIP–CO_2): comparative analysis of EMIC simulations of climate, and of equilibrium and transient responses to atmospheric CO_2 doubling. Clim Dynam 25(4):363-385

141. Petoukhov VK (1980) A zonal climate model of heat and moisture exchange in the atmosphere over the underlying layers of ocean and land. In: Golitsyn GS, Yaglom AM (eds) Physics of the atmosphere and the problem of climate, Nauka, Moscow. pp 8-41

142. Pitman AJ, de Noblet-Ducoudre N, Cruz FT, Davin EL, Bonan GB, Brovkin V, Claussen M, Delire C, Ganzeveld L, Gayler V, van den Hurk BJJM, Lawrence PJ, van der Molen MK, Muller C, Reick CH, Seneviratne SI, Strengers BJ, Voldoire A (2009) Uncertainties in climate responses to past land cover change: first results from the LUCID intercomparison study. Geophys Res Lett 36:L14814 doi:10.1029/2009gl039076

143. Plattner G-K, Knutti R, Joos F, Stocker T, Von Bloh W, Brovkin V, Cameron D, Driesschaert E, Dutkiewicz S, Eby M (2008) Long-term climate commitments projected with climate–carbon cycle models. J Climate 21:2721-2751

144. Plattner GK, Joos F, Stocker T, Marchal O (2001) Feedback mechanisms and sensitivities of ocean carbon uptake under global warming. Tellus B 53(5):564-592

145. Pongratz J, Reick CH, Raddatz T, Claussen M (2010) Biogeophysical versus biogeochemical climate response to historical anthropogenic land cover change. Geophys Res Lett 37:L08702 doi:10.1029/2010gl043010

146. Post WM, Peng T-H, Emanuel WR, King AW, Dale VH, DeAngelis DL (1990) The global carbon cycle. Am Sci 78(4):310-326

147. Raiswell R, Canfield DE (2012) The iron biogeochemical cycle past and present. Geochem Perspect 1(1):1-220 doi:10.7185/Geochempersp.1.1

148. Rasch PJ, Tilmes S, Turco RP, Robock A, Oman L, Chen C-CJ, Stenchikov GL, Garcia RR (2008) An overview of geoengineering of climate using stratospheric sulphate aerosols. Philos T R Soc A 366(1882):4007-4037

149. Reich PB, Hobbie SE (2013) Decade-long soil nitrogen constraint on the CO_2 fertilization of plant biomass. Nature Clim Change 3(3):278-282 doi:10.1038/nclimate1694

150. Reich PB, Hobbie SE, Lee T, Ellsworth DS, West JB, Tilman D, Knops JMH, Naeem S, Trost J (2006) Nitrogen limitation constrains sustainability of ecosystem response to CO_2. Nature 440(7086):922-925 doi:10.1038/nature04486

151. (2009) Geoengineering the climate: science, governance and uncertainty. Royal Society, London.

152. Rubasinghege G, Lentz RW, Scherer MM, Grassian VH (2010) Simulated atmospheric processing of iron oxyhydroxide minerals at low pH: roles of particle size and acid anion in iron dissolution. P Natl Acad Sci 107(15):6628-6633

153. Sabine CL, Feely RA (2007) The oceanic sink for carbon dioxide. In: Reay D, Hewitt N, Grace J, Smith K (eds) Greenhouse gas sinks, CABI Publishing, Oxfordshire. pp 31-49

154. Sabine CL, Feely RA, Gruber N, Key RM, Lee K, Bullister JL, Wanninkhof R, Wong CS, Wallace DW, Tilbrook B, Millero FJ, Peng TH, Kozyr A, Ono T, Rios AF (2004) The oceanic sink for anthropogenic CO_2. Science 305(5682):367-371 doi:10.1126/science.1097403

155. Saikawa E, Prinn RG, Dlugokencky E, Ishijima K, Dutton GS, Hall BD, Langenfelds R, Tohjima Y, Machida T, Manizza M, Rigby M, O'Doherty S, Patra PK, Harth CM, Weiss RF, Krummel PB, van der Schoot M, Fraser PB, Steele LP, Aoki S, Nakazawa T, Elkins JW (2013) Global and regional emissions estimates for N_2O. Atmos Chem Phys Discuss 13(7):19471-19525 doi:10.5194/acpd-13-19471-2013

156. Sakamoto TT, Komuro Y, Nishimura T, Ishii M, Tatebe H, Shiogama H, Hasegawa A, Toyoda T, Mori M, Suzuki T (2012)

MIROC4h—a new high-resolution atmosphere-ocean coupled general circulation model. J Meteorol Soc Jpn 90(3):325-359

157. Saltzman B (1978) A survey of statistical-dynamical models of the terrestrial climate. Adv Geophys 20:183-304

158. Santana-Casiano JM, González-Dávila M, Rueda MJ, Llinás O, González-Dávila EF (2007) The interannual variability of oceanic CO_2 parameters in the northeast Atlantic subtropical gyre at the ESTOC site. Global Biogeochem Cy 21(1):GB1015 doi:10.1029/2006GB002706

159. Sarmiento J, Slater R, Dunne J, Gnanadesikan A, Hiscock M (2010) Efficiency of small scale carbon mitigation by patch iron fertilization. Biogeosciences 7(11):3593-3624

160. Sarmiento JL, Orr JC (1991) Three-dimensional simulations of the impact of Southern Ocean nutrient depletion on atmospheric CO_2 and ocean chemistry. Limnol Oceanogr 36(8):1928-1950

161. Sarmiento JL, Orr JC, Siegenthaler U (1992) A perturbation simulation of CO_2 uptake in an ocean general-circulation model. J Geophys Res-Oceans 97(C3):3621-3645 doi:10.1029/91jc02849

162. Sato H, Ito A, Ito A, Ise T, Kato E (2014) Current status and future of land surface models. Soil Sci Plant Nutri, doi:10.1080/00380 768.2014.917593

163. Sedwick PN, Sholkovitz ER, Church TM (2007) Impact of anthropogenic combustion emissions on the fractional solubility of aerosol iron: Evidence from the Sargasso Sea. Geochem Geophy Geosy 8:Q10Q06 doi:10.1029/2007gc001586

164. Sellers WD (1969) A global climatic model based on the energy balance of the earth-atmosphere system. J Appl Meteorol 8(3):392-400

165. Shevliakova E, Pacala SW, Malyshev S, Hurtt GC, Milly P, Caspersen JP, Sentman LT, Fisk JP, Wirth C, Crevoisier C (2009) Carbon cycling under 300 years of land use change: Importance of the secondary vegetation sink. Global Biogeochem Cy 23:GB2022 doi:10.1029/2007GB003176

166. Shi Z, Bonneville S, Krom M, Carslaw K, Jickells T, Baker A, Benning LG (2011) Iron dissolution kinetics of mineral dust at low pH during simulated atmospheric processing. Atmos Chem Phys 11(3):995-1007

167. Shi Z, Krom MD, Bonneville S, Baker AR, Bristow C, Drake N, Mann G, Carslaw K, McQuaid JB, Jickells T (2011) Influence of chemical weathering and aging of iron oxides on the potential iron solubility of Saharan dust during simulated atmospheric processing. Global Biogeochem Cy 25:GB2010 doi:10.1029/2010GB003837

168. Shi Z, Krom MD, Jickells TD, Bonneville S, Carslaw KS, Mihalopoulos N, Baker AR, Benning LG (2012) Impacts on iron solubility in the mineral dust by processes in the source region and the atmosphere: a review. Aeolian Res 5:21-42

169. Shindell DT, Schmidt GA (2004) Southern Hemisphere climate response to ozone changes and greenhouse gas increases. Geophys Res Lett 31:L18209 doi:10.1029/2004GL020724

170. Sholkovitz ER, Sedwick PN, Church TM, Baker AR, Powell CF (2012) Fractional solubility of aerosol iron: synthesis of a global-scale data set. Geochim Cosmochim Ac 89:173-189

171. Siefert RL, Johansen AM, Hoffmann MR (1999) Chemical characterization of ambient aerosol collected during the southwest monsoon and intermonsoon seasons over the Arabian Sea: labile-Fe (II) and other trace metals. J Geophys Res 104(D):3511-3526

172. Smith R (2012) The FAMOUS climate model (versions XFXWB and XFHCC): description update to version XDBUA. Geosci Model Dev 5(1):269-276

173. Sokolov AP, Kicklighter DW, Melillo JM, Felzer BS, Schlosser CA, Cronin TW (2008) Consequences of considering carbon-nitrogen interactions on the feedbacks between climate and the terrestrial carbon cycle. J Climate 21(15):3776-3796

174. Solmon F, Chuang PY, Meskhidze N, Chen Y (2009) Acidic processing of mineral dust iron by anthropogenic compounds over the north Pacific Ocean. J Geophys Res-Atmospheres 114:D02305 doi:10.1029/2008jd010417

175. Steinacher M, Joos F, Frolicher TL, Plattner GK, Doney SC (2009) Imminent ocean acidification in the Arctic projected with the NCAR global coupled carbon cycle-climate model. Biogeosciences 6(4):515-533

176. Stocker BD, Roth R, Joos F, Spahni R, Steinacher M, Zaehle S, Bouwman L, Xu R, Prentice IC (2013) Multiple greenhouse-gas

feedbacks from the land biosphere under future climate change scenarios. Nature Clim Change 3(7):666-672 doi:10.1038/nclimate1864

177. Strong AL, Cullen JJ, Chisholm S (2009) Ocean fertilization: science, policy, and commerce. Oceanography 22:236-261

178. Tachiiri K, Hargreaves J, Annan J, Oka A, Abe-Ouchi A, Kawamiya M (2010) Development of a system emulating the global carbon cycle in Earth system models. Geosci Model Dev 3(2):365-376

179. Tachiiri K, Hargreaves JC, Annan JD, Huntingford C, Kawamiya M (2013) Allowable carbon emissions for medium-to-high mitigation scenarios. Tellus B 65:20586 doi:10.3402/tellusb.v65i0.20586

180. Tachiiri K, Ito A, Hajima T, Hargreaves JC, Annan JD, Kawamiya M (2012) Nonlinearity of land carbon sensitivities in climate change simulations. J Meteorol Soc Jpn 90:259-274

181. Takahashi T, Sutherland SC, Sweeney C, Poisson A, Metzl N, Tilbrook B, Bates N, Wanninkhof R, Feely RA, Sabine C (2002) Global sea–air CO_2 flux based on climatological surface ocean pCO_2, and seasonal biological and temperature effects. Deep Sea Res Part II: Top Stud Oceanograph 49(9):1601-1622

182. Takahashi T, Sutherland SC, Wanninkhof R, Sweeney C, Feely RA, Chipman DW, Hales B, Friederich G, Chavez F, Sabine C (2009) Climatological mean and decadal change in surface ocean pCO_2, and net sea–air CO_2 flux over the global oceans. Deep Sea Res Part II: Top Stud Oceanograph 56(8):554-577

183. Taylor KE, Stouffer RJ, Meehl GA (2012) An overview of CMIP5 and the experiment design. B Am Meteorol Soc 93(4):485-498

184. Thompson DW, Solomon S (2002) Interpretation of recent Southern Hemisphere climate change. Science 296(5569):895-899

185. Thornton PE, Doney SC, Lindsay K, Moore JK, Mahowald N, Randerson JT, Fung I, Lamarque JF, Feddema JJ, Lee YH (2009) Carbon-nitrogen interactions regulate climate-carbon cycle feedbacks: results from an atmosphere-ocean general circulation model. Biogeosciences 6(10):2099-2120

186. Timmermann A, Timm O, Stott L, Menviel L (2009) The roles of CO_2 and orbital forcing in driving southern hemispheric temperature

variations during the last 21000 Yr. J Climate 22:1626-1640 doi:10.1175/2008JCLI2161.

187. Tuenter E, Weber S, Hilgen F, Lourens L, Ganopolski A (2005) Simulation of climate phase lags in response to precession and obliquity forcing and the role of vegetation. Clim Dynam 24(2-3):279-295doi:0.1007/s00382-004-0490-1

188. Uematsu M (2013) Study on the chemical substance transported to the ocean through the atmosphere. Uminokenkyu (Oceanography in Japan) 22(2):35-45 (in Japanese)

189. Van Vuuren D, Meinshausen M, Plattner G-K, Joos F, Strassmann KM, Smith SJ, Wigley T, Raper S, Riahi K, De La Chesnaye F (2008) Temperature increase of 21st century mitigation scenarios. P Natl Acad Sci 105(40):15258-15262

190. Volk T, Hoffert MI (1985) Ocean carbon pumps: analysis of relative strengths and efficiencies in ocean-driven atmospheric CO_2 changes. Geoph Monog Series 32:99-110

191. Watanabe S, Hajima T, Sudo K, Nagashima T, Takemura T, Okajima H, Nozawa T, Kawase H, Abe M, Yokohata T, Ise T, Sato H, Kato E, Takata K, Emori S, Kawamiya M (2011) MIROC-ESM: model description and basic results of CMIP5-20c3m experiments. Geosci Model Dev 4:845-872

192. Weaver AJ, Eby M, Wiebe EC, Bitz CM, Duffy PB, Ewen TL, Fanning AF, Holland MM, MacFadyen A, Matthews HD (2001) The UVic Earth System Climate Model: model description, climatology, and applications to past, present and future climates. Atmos Ocean 39(4):361-428

193. Weaver AJ, Sedláček J, Eby M, Alexander K, Crespin E, Fichefet T, Philippon-Berthier G, Joos F, Kawamiya M, Matsumoto K, Steinacher M, Tachiiri K, Tokos K, Yoshimori M, Zickfeld K (2012) Stability of the Atlantic meridional overturning circulation: a model intercomparison. Geophys Res Lett 39:L20709 doi:10.1029/2012GL053763

194. Weber SL (2010) The utility of Earth system models of intermediate complexity (EMICs). Wiley Interdiscip Rev-Climate Change 1(2):243-252

195. Webster M, Sokolov AP, Reilly JM, Forest CE, Paltsev S, Schlosser A, Wang C, Kicklighter D, Sarofim M, Melillo J (2012) Analysis of climate policy targets under uncertainty. Climatic Change 112(3-4):569-583

196. Willey JD, Mullaugh KM, Kieber RJ, Avery GB Jr, Mead RN (2012) Controls on the redox potential of rainwater. Envir Sci Tech 46:13103-13111 doi:10.1021/ es302569j

197. Yamamoto A, Kawamiya M, Ishida A, Yamanaka Y, Watanabe S (2012) Impact of rapid sea-ice reduction in the Arctic Ocean on the rate of ocean acidification. Biogeosciences 9(6):2365-2375 doi:10.5194/Bg-9-2365-2012

198. Yao X, Fang M, Chan CK (2002) Size distributions and formation of dicarboxylic acids in atmospheric particles. Atmos Environ 36(13):2099-2107

199. Yoshikawa C, Kawamiya M, Kato T, Yamanaka Y, Matsuno T (2008) Geographical distribution of the feedback between future climate change and the carbon cycle. J Geophys Res 113(G3):G03002 doi:10.1029/2007jg000570

200. Yu JZ, Huang X-F, Xu J, Hu M (2005) When aerosol sulfate goes up, so does oxalate: implication for the formation mechanisms of oxalate. Envir Sci Tech 39(1):128-133

201. Zaehle S, Ciais P, Friend AD, Prieur V (2011) Carbon benefits of anthropogenic reactive nitrogen offset by nitrous oxide emissions. Nat Geosci 4(9):601-605 doi:10.1038/ngeo1207

202. Zaehle S, Friedlingstein P, Friend AD (2010) Terrestrial nitrogen feedbacks may accelerate future climate change. Geophys Res Lett 37:L01401 doi:10.1029/2009gl041345

203. Zhuang G, Yi Z, Duce RA, Brown PR (1992) Link between iron and sulphur cycles suggested by detection of Fe(n) in remote marine aerosols. Nature 355:537-539

204. Zickfeld K, Eby M, Matthews HD, Schmittner A, Weaver AJ (2011) Nonlinearity of carbon cycle feedbacks. J Climate 24(16):4255-4275

205. Zickfeld K, Eby M, Weaver AJ, Alexander K, Crespin E, Edwards NR, Eliseev AV, Feulner G, Fichefet T, Forest CE, Friedlingstein P, Goosse H, Holden PB, Joos F, Kawamiya M, Kicklighter D, Kienert H, Matsumoto K, Mokhov II, Mokhov E, Olsen SM, Pedersen JOP, Perrette M, Philippon-Berthier G, Ridgwell A, Schlosser A, Schneider Von Deimling T, Shaffer G, Sokolov A, Spahni R, et al. (2013) Long-term climate change commitment and reversibility: an EMIC intercomparison. J Climate 26(16):5782-5809

On the Average Temperature of Airless Spherical Bodies and the Magnitude of Earth's Atmospheric Thermal Effect

Den Volokin and Lark ReLlez

Tso Consulting, 843 E Three Fountains Suite 260, Salt Lake City, UT 84107, USA

ABSTRACT

The presence of atmosphere can appreciably warm a planet's surface above the temperature of an airless environment. Known as a natural Greenhouse Effect (GE), this near-surface Atmospheric Thermal Enhancement (ATE) as named herein is presently entirely attributed to the absorption of up-welling long-wave radiation by greenhouse gases. Often quoted as 33 K for Earth, GE is estimated as a difference between planets's observed mean surface temperature and an effective radiating temperature calculated from the globally averaged absorbed solar flux using the Stefan-Boltzmann (SB) radiation law. This approach equates a planet's average temperature in the absence of greenhouse gases or atmosphere to an effective emission temperature assuming $ATE \equiv GE$. The SB law is also routinely employed to estimating the mean temperatures of airless bodies. We demonstrate that this formula as applied to spherical objects is mathematically incorrect owing to Hölder's inequality between integrals and leads to biased results such as a significant underestimation of Earth's ATE. We derive a new expression for the mean physical temperature of airless bodies based on an analytic integration of the SB law over a sphere that accounts for effects of regolith heat storage and cosmic background radiation on nighttime temperatures. Upon verifying our model against Moon surface temperature data provided by the NASA Diviner Lunar Radiometer Experiment, we propose it as a new analytic standard for evaluating the thermal environment of airless bodies. Physical evidence is presented that Earth's ATE should be assessed against the temperature of an equivalent airless body such as the Moon rather than a hypothetical atmosphere devoid of greenhouse gases. Employing the new temperature formula we show that Earth's *total* ATE is ~90 K, not 33 K, and that $ATE = GE + TE$, where GE is the thermal effect of greenhouse gases, while $TE > 15$ K is a thermodynamic enhancement independent of the atmospheric infrared back radiation. It is concluded that the contribution of greenhouse gases to Earth's ATE defined as $GE = ATE - TE$ might be greater than 33 K, but will remain uncertain until the strength of the hereto identified TE is fully quantified by future research.

BACKGROUND

It is an undisputed fact that the atmosphere can appreciably heat a planet's surface above the temperature of an airless environment receiving the same stellar irradiance. Known as a natural Greenhouse Effect (GE), this extra atmospheric warmth is presently completely attributed to the absorption and re-emission of upwelling long-wave radiation by heat-absorbing gases such as CO_2, water vapor, methane (CH_4), nitrous oxide (N_2O) and others (Schmidt et al. 2010; Lacis et al. 2010). Thus, GE has two scientific measures at the present (Lacis et al. 2013): a) as an observed difference in the outgoing global infrared flux (W m^{-2}) between the planet surface and the top of the atmosphere (Ramanathan and Inamdar 2006; Schmidt et al. 2010; Pierrehumbert 2011); and b) as an extra warmth or increased temperature at the surface (Hansen et al. 1981; Schmidt et al. 2010; Lacis et al.2010, 2013). This study explores the latter measure of GE using Earth as an example. The additional warmth provided by GE creates climate conditions that foster life on our Planet by enabling the existence of liquid oceans and providing for a global water cycle (Pierrehumbert 2010). In order to better distinguish between the two measures of GE and to facilitate a proper understanding of our analysis and results, we hereto introduce the term *Atmospheric Thermal Enhancement* (ATE) to describe the *total* extra warmth near a planet surface measured as a difference (K) between the planet's present mean global surface temperature and an estimated planetary reference temperature in the *absence* of atmosphere. By referring to the whole atmosphere, ATE also allows for investigation of potential contributions beyond those currently attributed to greenhouse gases.

According to satellite observations, Earth's atmosphere retains on average 155–158 W m^{-2} of the upwelling long-wave radiation emitted by the surface (Kiehl and Trenberth 1997; Trenberth et al.2009; Stephens et al. 2012; Wild et al. 2013). This infrared heat absorption by greenhouse gases a.k.a. long-wave radiative forcing (Kiehl and Trenberth 1997) is presently believed to drive 100% of the near-surface ATE (Peixoto and Oort 1992; Lacis et al. 2010; Pierrehumbert 2010; Schmidt et al.2010). Most researchers assume that greenhouse gases boost the Earth's mean global surface temperature by about 33 K (e.g. Hansen et al. 1981; Peixoto and Oort 1992; Wallace and Hobbs2006; Lacis et al. 2010, 2013; Schmidt et al. 2010). Some argue that Earth's GE is only ~20 K (e.g. Zeng 2010). Knowing the exact magnitude of this natural atmospheric effect is important because it might relate to the planet's long-term climate sensitivity to anthropogenic greenhouse emissions. The goal of this study is to examine the current method for calculating the thermal effect of planetary atmospheres and reassess the magnitude of Earth's ATE as an example using a new approach to estimating the average global temperature of *airless* celestial bodies validated against recent NASA observations and model simulations of the Moon thermo-physical environment.

Stefan-Boltzmann Radiation Law

According to the Stefan-Boltzmann (SB) law, any physical object with a temperature above the absolute zero emits radiation with a total intensity that is proportional to the 4th power of the object's absolute surface temperature. This implies that an object's equilibrium surface temperature (T, K) can be calculated from the amount of absorbed radiation (I, W m^{-2}) using the relation

$$T = \left(\frac{I}{\epsilon \sigma} \right)^{\frac{1}{4}}$$

(1)

Where ε is the object's broadband thermal emissivity/absorptivity ($0 \leq \varepsilon \leq 1$) and $\sigma = 5.6704 \times 10^{-8}$ W m^{-2} K^{-4} is the SB constant. A theoretical blackbody has $\varepsilon = 1.0$, while the emissivity of real objects such as soil and regolith is typically in the range $0.95 \leq \varepsilon \leq 0.99$ for far infrared wavelengths. A key assumption of Eq. (1) is that the object has an isothermal surface, which absorbs and emits a spatially homogeneous flux of radiation I.

Current Application of the SB Law to Planetary Bodies

The absorption of solar radiation by a spherical body varies with latitude and the time of day as a function of the solar incidence angle and the local surface albedo. However, the globally averaged flux of absorbed solar radiation S_a (W m^{-2}) can reliably be calculated using the formula:

$$S_a = \frac{S_o}{4} \left(1 - \alpha_p \right)$$

(2)

Where S_o is the solar irradiance (W m^{-2}), i.e. the flux incident on a plane perpendicular to the solar rays at the top of the atmosphere

(TOA), and α_p is the planetary Bond albedo (decimal fraction). The factor 1/4 serves to distribute the solar flux from a flat surface to a sphere and arises from the fact that the surface area of a sphere is 4 times larger than the surface area of a disk with the same radius. Hence, it seems logical that one could calculate a global equilibrium temperature for a planet from S_a using the SB law, i.e.

$$T_e = \left(\frac{S_a}{\epsilon \, \sigma} \right)^{1/4} = \left[\frac{S_o \left(1 - \alpha_p \right)}{4 \, \epsilon \, \sigma} \right]^{1/4}$$

(3)

In this expression, T_e is known as 'effective-emission' or 'radiating equilibrium' temperature (K), since it corresponds to the globally averaged radiation flux absorbed by a celestial body. Hereafter, we use the term effective emission temperature to denote quantities calculated from spatially averaged fluxes of absorbed solar radiation. This is in contrast to other terms we use such as 'mean physical', 'average equilibrium', or 'average skin' temperature that refer to quantities obtained via area-weighted averaging of observable or measured surface temperatures.

Equations (2) and (3) were first introduced to planetary science in the early 1960s (Blanco and McCuskey 1961; Möller 1964) and have been utilized ever since to estimate the average global temperatures of airless or nearly airless celestial bodies such as Mercury, Moon and Mars (e.g. Williams2014), to quantify the strength of greenhouse effects of planetary atmospheres (e.g. Hansen et al.1981; Lacis et al. 2013), and to determine the boundaries of Habitable Zones around stars (e.g. Kaltenegger and Sasselov 2011; Schulze-Makuch et al. 2011).

Employing typical values for Earth, i.e. $S_o = 1,360.9$ W m^{-2} (Kopp and Lean 2011), $\alpha_p = 0.294$ (Loeb et al. 2009; Stephens et al. 2012; Wild et al. 2013) and assuming $\varepsilon = 1.0$, formulas (2) and (3) yield $S_a = 240.2$ W m^{-2} and $T_e = 255.1$ K, respectively. The latter estimate is the basis for the frequently quoted 255 K (−18 C) mean global temperature of Earth in the absence of GE, i.e. if the Earth's atmosphere were absent or completely transparent to the outgoing infrared radiation (e.g.

Pierrehumbert 2010). According to the NOAA National Climatic Data Center, Earth's observed mean surface temperature (T_s) has been stable over the past 16 years and equals 287.6 K (+14.47 C). Thus, the current method quantifies GE as $T_s - T_e = 287.6 - 255.1 = 32.5$ K. Most studies assume a planetary albedo of 0.3 and arrive at GE $\approx$ 33 K. The present Greenhouse theory attributes Earth's entire atmospheric thermal effect to the absorption and re-emission of outgoing long-wave radiation by tropospheric greenhouse gases assuming ATE $\equiv$ GE (Hansen et al. 1981; Peixoto and Oort 1992; Wallace and Hobbs 2006; Marshall and Plumb 2008; Pierrehumbert 2010; Schmidt et al. 2010; Schulze-Makuch et al. 2011; Lacis et al. 2010, 2013).

Some authors (e.g. Zeng 2010) argue that the 33 K GE estimate rests on a logical caveat, since it is based on a reference temperature computed from Eq. (3) using Earth's full albedo $\alpha_p = 0.3$ that includes the radiative effects of clouds and water vapor. In order for a temperature to be able to serve as a proper reference in this case it must describe the planet's surface thermal environment in the *absence* of greenhouse gases. Removing heat-absorbing gases from Earth's atmosphere, of which water vapor is primary (Schmidt et al. 2010; Lacis et al. 2013), would reduce the Earth albedo well below 0.294, since the scattering of sunlight by clouds and airborne water molecules accounts for about 50% of the planet's total shortwave reflectance. Hence, quantifying the strength of GE logically requires using a surface albedo (α_o) in Eq. (3) that is free from the radiative effects of atmospheric water (Zeng 2010). Following a similar logic, we argue that Earth's *total* ATE ought to be evaluated against the temperature of an equivalent *airless* body rather than a hypothetical atmosphere devoid of greenhouse gases. This is because, in addition to vapor clouds, air molecules and airborne aerosols significantly contribute to the atmospheric albedo as well as for other reasons related to the planet's surface thermal conductivity explained below.

Recent analyses of Earth's global energy budget based on satellite observations (Stephens et al. 2012) and ground measurements (Wild et al. 2013) suggest $0.122 \leq \alpha_o \leq 0.13$ for the Earth averaged land-ocean albedo. Serendipitously, these values are similar to the Moon's 0.136 average broadband albedo measured by the Clouds and the Earth's Radiant Energy System (CERES) (Matthews 2008) and the 0.131 effective lunar-regolith albedo estimated in this study (see discussion below). Using the satellite-observed value $\alpha_o = 0.122$

in Eq. (3) produces $T_e = 269.4$ K for Earth, which translates into ATE $\equiv$ GE $= 287.6 - 269.4 = 18.2$ K according to the present method based on the effective emission temperature. Zeng (2010) arrived at ATE ≈ 20 K by assuming a somewhat higher Earth surface albedo $\alpha_o = 0.14$. We concur that, in the context of Eq. (3), the 18–20 K estimate of ATE is theoretically more justifiable than the canonic 33 K value obtained by employing Earth's total albedo. It is important to note that all popular estimates of the atmospheric thermal effect ranging from 18 K to 33 K are based on Eq. (3) or similar 1-D radiative-transfer models and were not derived from 3-D global circulation models.

The above discussion makes it clear that quantifying the *total* magnitude of ATE requires an accurate estimation of the planet's equilibrium mean surface temperature in the *absence* of atmosphere. In general, we hereto refer to such a 'no-atmosphere' estimate as the average skin temperature (T_{na}) of an Airless Spherical Celestial Object (ASCO). Obviously, T_{na} depends on solar irradiance and the surface albedo, and ATE $= T_s - T_{na}$. Current climate and planetary sciences oftentimes identify mean physical temperatures of airless celestial bodies with their effective emission temperatures implicitly assuming $T_s \equiv T_{na}$ (e.g. Schulze-Makuch et al. 2011; Lacis et al. 2013). For example, the average global temperatures reported by the NASA Planetary Factsheet (Williams 2014) for the Moon (270.7 K), Mercury (440 K) and even Mars (210 K) have been calculated from Eq. (3). However, there is a theoretical problem with this formula as applied to spherical bodies related to what is known in mathematics as Hölder's inequality between integrals (Beckenbach and Bellman 1983; Abualrub and Sulaiman 2009). The problem has been identified by previous research (e.g. Leconte et al. 2013), but it has not been thoroughly analyzed in terms of its implications for the physical meaning and usefulness of T_e.

Hölder's Inequality and its Implications for Planetary Flux-temperature Relationships

In its general form, Hölder's inequality states that, for any pair of measurable real- or complex-valued functions $f(x)$ and $g(x)$, the following relationship is always true

$$\int |f(x)\,g(x)|\,dx \;\leq\; \left\{ \int |f(x)|^p dx \right\}^{\frac{1}{p}} \left\{ \int |g(x)|^q dx \right\}^{\frac{1}{q}}$$

(4a)

Provided $1 \leq p, q < \infty$ and $1/p + 1/q = 1$ (Beckenbach and Bellman 1983). In regard to the SB law and the latitudinal distribution of equilibrium temperatures $T(\mu)$ on the surface of a sphere (where $0 \leq \mu \leq 1$ is an area-weighting factor defined as the cosine of latitude), the relevant form of Hölder's inequality is obtained from (4a) by substituting $f(x) = T(\mu)$, $g(x) = 1$, $p = 4$ and $q = 4/3$. This produces:

$$\int_0^1 T(\mu)\,d\mu \;<\; \left\{ \int_0^1 T(\mu)^4\,d\mu \right\}^{\frac{1}{4}}$$

(4b)

Inequality (4b) implies that the area-weighted average temperature of a spherical surface (on the left) is always *lower* than the temperature calculated from the area-weighted average long-wave radiation emitted by the surface in proportion to $T(\mu)^4$ (on the right). Due to a non-linear relationship between temperature and the emitted radiative bolometric flux, and a strong latitudinal variation of the absorbed shortwave radiation across the surface of a sphere, the actual mean global temperature of a directionally illuminated planet is *not* estimable in principle from a planetary averaged radiative flux (Eq. 2) using the SB law (Eq. 1). This is because a spherical geometry violates the fundamental assumption in the SB relationship for spatial homogeneity of radiation absorption and emission. Hence, Eq. (3) yields the temperature of a flat isothermal surface rather than the average temperature of a thermally heterogeneous sphere as required for planets. In other words, T_e is the equilibrium temperature of a *black disk* orthogonally illuminated by shortwave radiation with intensity equal to the average solar flux absorbed by a sphere having a Bond albedo α_p. This makes T_e a *non-physical* temperature with respect to a spherical surface. The effect of Hölder's inequality can be illustrated with the following example.

Consider two points, P_1 and P_2, on the surface of an ASCO located at the exact same latitude (e.g. 45°N) but at opposite longitudes so that, when P_1 is fully illuminated, P_2 is completely shaded and vice versa (Figure 1). If such an ASCO orbited the Sun at Earth's distance, had a regolith of zero thermal conductivity, and were only heated by solar radiation, then the equilibrium temperature of the illuminated point would be $T_1 = [S_o(1-a_o)\ \cos\theta/\varepsilon\sigma]^{0.25} = 349.6$ K assuming $a_o = 0.12$ (a typical value for rocky surfaces), a solar incident angle $\theta = 45°$, and $\varepsilon = 1.0$. The temperature of the shaded point would be $T_2 = 0$, because it receives no radiation since $\cos\theta < 0$ and there is no heat release from the regolith at night due to zero heat storage. The mean physical temperature between the two points is simply then $T_m = (T_1 = T_2)/2 = 174.8$ K. However, if one employs the average solar flux absorbed between the two points, i.e. $S_m = \{[S_o(1-a_o)\cos\theta] + 0\}/2 = 423.4$ W m^{-2} to calculate a 'mean' effective emission temperature, one obtains $T_e = [S_m/\varepsilon\sigma]^{0.25} = 294.0$ K. Clearly $T_e \gg T_m$, a result of Hölder's inequality.

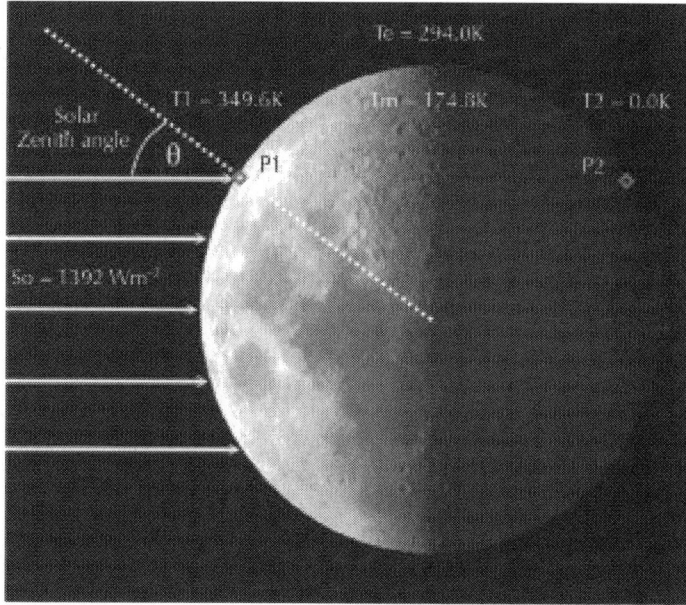

Figure 1: Illustration of Hölder's inequality between integrals. Due to a non-linearity of the SB law and a non-uniform distribution of the incident solar radiation on the surface of a sphere, the equilibrium temperature (T_e) computed

from a spatially averaged radiation flux is always higher than the arithmetic average temperature (T_m).

The conclusion from the above discussion is that a proper calculation of the mean *physical* temperature of an airless celestial body (T_{na}) requires an explicit integration of the SB law over the planet surface. This means *first* taking the 4th root of the absorbed shortwave flux at every point on the planet and *then* averaging the resulting temperature field across the entire surface rather than calculating a single temperature from the globally averaged absorbed solar flux as done in Eq. (3). It should be pointed out that global climate models intrinsically account for Hölder's inequality by virtue of being three-dimensional and explicitly resolving the spatial heterogeneity of radiation absorption and emission (as well as other energy transport processes) within the context of a spherical geometry. However, 3-D models have not historically been applied to assess the strength of Earth's ATE (GE). Hence, our critique is strictly directed towards the effective emission-temperature formula (3) and other similar 1-D radiative-transfer models (e.g. Manabe and Möller 1961; Manabe and Strickler 1964).

From the standpoint of Hölder's inequality, one would expect T_{na} to approach T_e only if the absorbed solar radiation were uniformly distributed throughout the entire planet surface. However, this requires a regolith of *infinite* lateral thermal conductivity, which is physically impossible. Real ASCOs such as the Moon have extremely low surface thermal conductivities (Vasavada et al. 2012), and the absorbed solar flux varies greatly with latitude and solar angle resulting in a highly non-uniform distribution of surface temperatures. Hence, in the general case, we expect $T_{na} \ll T_e$. This implies that effective emission temperatures are not equivalent to and should not be confused with actual physical temperatures on a sphere, a conclusion also reached by Leconte et al. (2013). Some researchers identify Earth's $T_e \approx 255$ K with the observed average temperature at about 5 km altitude in the free troposphere (e.g. Hansen et al. 1981; Marshall and Plumb 2008; Pierrehumbert 2010). Others relate T_e of airless bodies to brightness temperatures retrieved via radio waves for ~1 m depth below the surface (e.g. Lissauer and Pater 2013, Chapter 4.1). However, Hölder's inequality reveals that the effective emission temperature of a spherical object is a mathematical abstraction with no physical analogue; hence, any numerical similarity between T_e and actual planetary temperatures measured at, below or above the surface must be viewed as a coincidence. Consequently, all estimates of GE (ATE) based on Eq. (3) are misleading, since they are products of comparisons between Earth's *observed* average surface air temperature (T_s) and some *unmeasurable* (non-physical) effective radiating temperatures (T_e) at the TOA. A proper assessment of ATE requires a reliable estimate of the planet's mean global *surface* temperature in the *absence* of atmosphere. Thus, there is a practical need for a new analytic model that accounts for Hölder's inequality while accurately predicting the average physical temperatures of airless spherical bodies.

METHODS

Derivation of an Analytic Model for the Mean Physical Temperature of Airless Bodies

In order to derive a formula for T_{na} that conforms to Hölder's inequality, we adopted the following reasoning. The equilibrium temperature T_i at a point i on the surface of an airless planet is determined by the incident solar flux and the local surface albedo during daytime, and by the upward heat flux emanating from the regolith at night. The nighttime release of heat from the ground is assumed to primarily originate from stored solar energy in the regolith with negligible contribution by geothermal sources (e.g. Vasavada et al. 2012). Hence, the nighttime heat flux can be approximated as a *fraction* of the solar radiation absorbed by the surface during daylight hours. A robust physical model of the average surface temperature of airless bodies must also include the small effect of geothermal fluxes and cosmic microwave background radiation (CMBR). The latter only becomes important under a low solar irradiance, i.e. for ASCOs orbiting at the outskirts of the solar system. Assuming a spatial uniformity of the sub-solar (normal) albedo across the planet surface and allowing the point albedo to vary with solar incidence angle, we can write the following general equation for T_i using the SB law:

$$T_i = \begin{cases} \left[\dfrac{(1-\eta)\,S_o\,[1-A(\theta_i)]\cos\theta_i + (R_C + R_g)}{\epsilon\,\sigma}\right]^{1/4} & \text{if } 0 \leq \theta_i < \dfrac{\pi}{2} \\[2em] \left[\dfrac{(R_C + R_g)-\eta\,S_o\,[1-A(\theta_i)]\cos\theta_i}{\epsilon\,\sigma}\right]^{1/4} & \text{if } \dfrac{\pi}{2} \leq \theta_i \leq \pi \end{cases}$$

$$(5)$$

Here, S_o is the solar (stellar) irradiance (W m^{-2}), θ_i is the incidence angle of shortwave radiation (rad) at point i (i.e. the angle between stellar rays and an axis normal to the surface at that point), $A(\theta_i)$ is the albedo as a function of θ_i, η is the fraction of absorbed solar flux stored into regolith through heat conduction, $R_c = \sigma\,2.725^4 = 3.13 \times 10^{-6}$ W m^{-2} is CMBR (Fixsen 2009), R_g is a spatially uniform geothermal flux (W m^{-2}), and ε is the average regolith long-wave emissivity; typically $0.95 \leq \varepsilon < 0.99$; in this study $\varepsilon = 0.98$. The upper portion of Eq. (5) describes the surface temperature at point i during daytime, while the lower portion defines the respective temperature at night. During daylight hours when the Sun is above the horizon ($0 \leq \theta_i \leq \pi/2$), the solar radiation absorbed at point i is partitioned into a flux giving the surface its daytime temperature, i.e. $(1-\eta)\,So\,[1-A\,(\theta_i)]\cos\theta_i$, and another smaller flux conducted and stored into the ground as heat. At night, when the Sun is below the horizon ($\pi/2 \leq \theta_i \leq \pi$), the stored heat is released giving the surface point its nighttime temperature; hence, the presence of the term $\eta\,So\,[1-A\,(\theta_i)]\cos\theta_i$ in the nighttime portion of Eq. (5). Since CMBR is virtually isotropic, we assume R_c to be uniformly absorbed by the daytime and nighttime hemisphere of an airless body. Note that Eq. (5) only describes location-specific differences in point equilibrium temperatures and does not simulate temporal temperature changes. This is so, because our objective is to conduct spatial integration and derive a *spherical* temperature average.

The Moon is our serendipitous ASCO example. In situ lunar measurements by the Apollo Mission and remote observations by the NASA Diviner Lunar Radiometer Experiment (DLRE) suggest that the albedo of regolith-covered surfaces in a vacuum varies with solar incidence angle according to the function (Keihm 1984; Vasavada et al. 2012):

$$A(\theta_i) = A_o + 0.045(\theta_i/45)^3 + 0.14(\theta_i/90)^8$$

(6)

Where θ_i is in degree and A_o is the normal (sub-solar) albedo at $\theta_i = 0°$. Figure 2 depicts the response of $A(\theta_i)$ to variations in the solar angle assuming $A_o = 0.105$, the average normal albedo suggested by Diviner equatorial measurements (Vasavada et al. 2012).

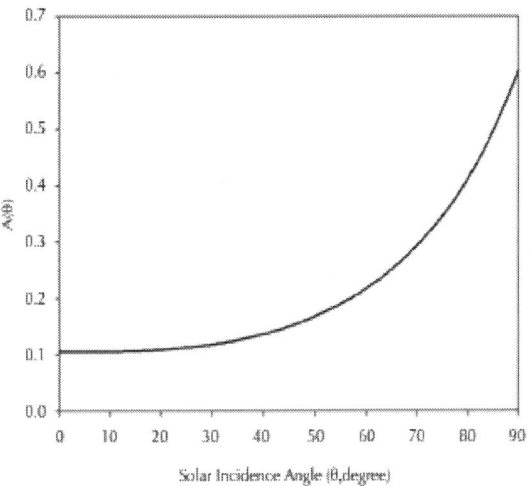

Figure 2: Variation of the moon's regolith albedo $A(\theta)$ as a function of solar incidence angle according to Eq. (6) based on surface measurements by Apollo and Diviner missions (Keihm 1984; Vasavada et al.2012).

Upon substituting $\mu \equiv \cos\theta_i$ in Eq. (5), the average global surface temperature of an airless celestial body T_{na} (K) is obtained from the spherical integration of Ti, i.e.

$$T_{na} = \frac{1}{4\pi} \int_0^{2\pi} \left(\int_{-1}^{1} T_i \, d\mu \right) d\varphi$$

(7)

Inserting the albedo function (6) into Eq. (5), however, renders integral (7) without a closed-form solution. To resolve this we plotted the absorption term $[1 - A(\theta_i)]\cos\theta_i$ in Eq. (5) versus the integration variable $\cos\theta_i$. The result is a monotonic relationship that can closely be approximated by a linear regression forced through the origin (Figure 3), i.e.

$$[1 - A(\theta_i)]\cos\theta_i \approx s_\theta \cos\theta_i$$

(8)

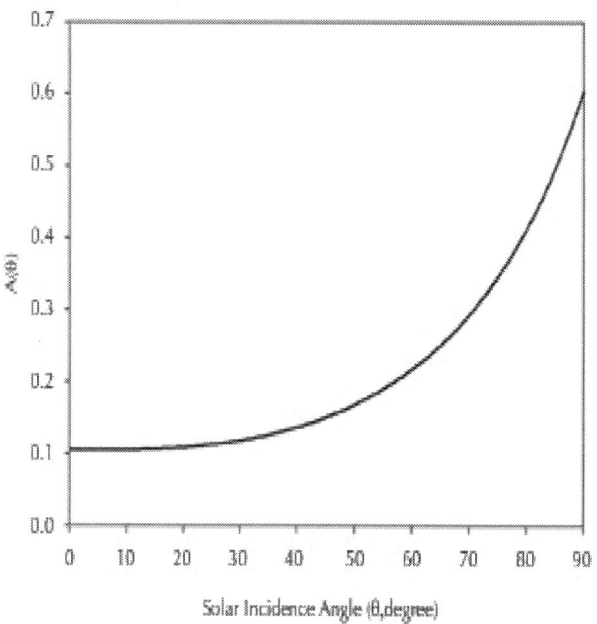

Figure 3: Emergent linear relationship between $\cos\theta$ and the regolith short-wave absorption term $[1 - A(\theta)]\cos\theta$, where θ is the solar incidence angle and $A(\theta)$ is the surface albedo as a function of θ defined by Eq. (6) (see Figure2).

With s_θ being the regression slope. Equation (8) implies $(1 - \alpha_e) = s_\theta$, where α_e is an *effective* surface albedo that incorporates the impact of a variable $A(\theta_i)$ on the surface temperature in the context of Eq. (5) and its spherical integral (7). Further analysis employing a range of values for A_o in Eq. (6) reveals:

$$\alpha_e \approx A_o + 0.026$$

(9)

Equation (9) yields $\alpha_e = 0.131$ for the Moon according to Diviner observations. As discussed below, the heat storage fraction η also varies with latitude. Thus, as with the albedo, it is more appropriate to use an *effective* heat storage fraction (η_e) in Eq. (5) rather than η.

The above transformations allow us to employ a fixed albedo (α_e) in Eq. (5) and solve integral (7) analytically to obtain a closed-form expression for T_{na}, i.e.

$$
\begin{aligned}
T_{na} &= \frac{1}{4\pi} \int_0^{2\pi} \left\{ \int_0^1 \left[\frac{(1-\eta_e)S_o(1-\alpha_e)\mu + R_c + R_g}{\epsilon\sigma} \right]^{1/4} d\mu \right. \\
&\quad \left. + \int_{-1}^0 \left[\frac{R_c + R_g - \eta_e S_o(1-\alpha_e)\mu}{\epsilon\sigma} \right]^{1/4} d\mu \right\} d\varphi \\
&= \frac{1}{4\pi} \int_0^{2\pi} \left\{ \frac{4}{5} \frac{\left[(1-\eta_e)S_o(1-\alpha_e) + R_c + R_g\right]^{5/4} - (R_C + R_g)^{5/4}}{(1-\eta_e)S_o(1-\alpha_e)(\epsilon\sigma)^{1/4}} \right. \\
&\quad \left. + \frac{4}{5} \frac{\left[\eta_e S_o(1-\alpha_e) + R_c + R_g\right]^{5/4} - (R_C + R_g)^{5/4}}{\eta_e S_o(1-\alpha_e)(\epsilon\sigma)^{1/4}} \right\} d\varphi \\
&= \frac{2}{5} \left\{ \frac{\left[(1-\eta_e)S_o(1-\alpha_e) + R_c + R_g\right]^{5/4} - (R_C + R_g)^{5/4}}{(1-\eta_e)S_o(1-\alpha_e)(\epsilon\sigma)^{1/4}} \right. \\
&\quad \left. + \frac{\left[\eta_e S_o(1-\alpha_e) + R_c + R_g\right]^{5/4} - (R_C + R_g)^{5/4}}{\eta_e S_o(1-\alpha_e)(\epsilon\sigma)^{1/4}} \right\}
\end{aligned}
$$

(10)

A numerical analysis of the final equation (10) reveals that the effect of CMBR on T_{na} is negligible for $S_o > 0.15$ W m^{-2}. In addition, the impact of geothermal fluxes on the surface temperature of airless bodies is oftentimes insignificant. Thus, in most cases, the above formula can be simplified by substituting $R_c = R_g = 0$. This produces:

$$T_{na} = \frac{2}{5} \left[\frac{S_o (1 - \alpha_e)}{\epsilon \sigma} \right]^{0.25} \Phi(\eta_e)$$

(11a)

Where $\Phi(\eta_e) \geq 1.0$ is given by:

$$\Phi(\eta_e) = (1 - \eta_e)^{0.25} + \eta_e^{0.25}$$

(11b)

The complete formula (10) only needs to be used if $S_o \leq 0.15$ W m^{-2} and/or R_g is significant compared to S_o. This is because, as $S_o \to 0.0$, Eq. (11a) approaches 0.0 as well, while Eq. (10) approaches 2.73 K, the irreducible minimum temperature of deep Space assuming $\varepsilon = 0.98$.

Conceptually $\Phi(\eta_e)$ is a non-dimensional thermal enhancement factor that boosts the global temperature of an airless planet above the level expected from a surface with zero thermal inertia, i.e. if the planet were completely non-conductive to heat. Thanks to $\eta_e > 0$, the night sides of rotating ASCOs remain at a significantly higher temperature than expected from CMBR alone. This substantially raises the global average temperature of ASCOs compared to the case when $\eta_e = 0$. In theory, η_e can vary in the interval $0.0 \leq \eta_e \leq 1.0$. However, due to physical constraints imposed by the low thermal conductivity of regolith in an airless environment, this range is considerably narrower in reality. For actual ASCOs, we expect $0.005 < \eta_e < 0.02$ based on thermal conductivity data for the lunar regolith reported by Vasavada et al. (2012). Figure 4a illustrates the response of $\Phi(\eta_e)$ to variation in η_e over the entire theoretical range, while Figure 4b depicts the same response over the approximate physically feasible range $0.0 \leq \eta_e \leq 0.02$. According to Eq. (11b), $\Phi(\eta_e)$ reaches a maximum of 1.682

at $\eta e = 0.5$. However, since it is not possible for a regolith immersed in vacuum to store on average as much as 50% of the absorbed solar energy as heat, $\Phi (\eta e)$ cannot practically ever reach its theoretical maximum. Realistically, we expect $1.26 < \Phi (\eta e) < 1.37$ for ASCOs.

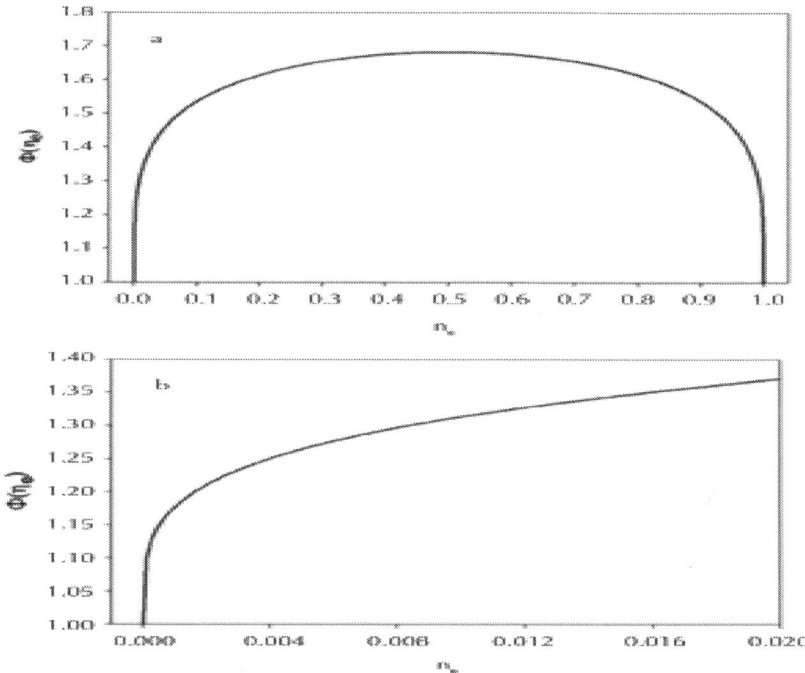

Figure 4: The planet thermal enhancement factor (Φ) as a function of the effective regolith heat storage fraction (η_e) according to Eq. (11b): a) over the entire theoretical range of η_e; b) over the physically feasible range of η_e.

Equation (11a) is similar to Eq. (15) in Rubincam (2004) and to Equation four in Leconte et al. (2013) except for the additional temperature enhancement factor $\Phi (\eta_e)$ that was not considered by these researchers. Previous studies have also not addressed the physical incompatibility between ASCO's actual average surface temperature and its effective emission temperature computed from Eq. (3). This incompatibility is revealed in our derivation by the following comparison. Using $S_o = 1,360.9$ W m^{-2}, $a_e = 0.131$ and $\varepsilon = 0.98$ in Eq. (3) yields an effective emission temperature for the Moon $T_e = 270.1$ K.

This estimate is 13.2 K *higher* than the theoretically maximum possible temperature $T_{na} = 256.9$ K produced by Eq. (11a) using the same input and a physically unreachable peak value of $\Phi(\eta) = 1.682$ corresponding to $\eta_e = 0.5$. Thus, in the absence of a significant geothermal flux, it is *in principle* not possible for an airless body to reach an average global temperature as high as its effective emission temperature.

Verification of the New Analytic Temperature Model for Airless Bodies

Since Eq. (10) and its simplified form (11a) were derived to predict the average surface temperature of celestial bodies with no tangible atmospheres, it is prudent to verify them against data for the Moon as the closest and best-studied airless object in the Solar System. Lunar temperatures have been measured for more than 50 years both remotely via Earth-based telescopes and instruments aboard lunar orbiters, and in situ by the Surveyor and Apollo landing mission (Paige et al. 2010a). Recently, the Diviner Lunar Radiometer Experiment (Paige et al. 2010a), a part of the Lunar Reconnaissance Orbiter (LRO) Mission (Vondrak et al. 2010), launched an extensive remote-sensing survey of the lunar surface. The Diviner instrument aboard LRO provides measurements in two spectral channels of reflected shortwave radiation and seven channels of emitted infrared long-wave radiation (Paige et al.2010a; Vasavada et al. 2012). The goal of DLRE is to map the Moon's surface temperature and albedo at a high spatial and temporal resolution over multiple diurnal and seasonal cycles. DLRE is the most comprehensive attempt to date to quantify the spatial and temporal variability of the lunar surface temperature. Although the project is still in progress, data acquired since the beginning of DLRE's commissioning period (the summer of 2009) already cover most of the Moon surface. These high-quality radiance measurements have been utilized to study thermal environments at the lunar equator, mid-latitudes and the Polar Regions, and to validate and refine existing thermo-physical models (Paige et al. 2010b; Bandfield et al. 2011; Vasavada et al. 2012).

Verification Approach

In order to obtain an independent estimate of the Moon's mean surface temperature needed to verify Equations (10) and (11a) we employed a detailed NASA thermo-physical model of the regolith called TWO that has been previously verified against Apollo multi-year borehole measurements on the Moon and remote-sensing observations of Mercury (Vasavada et al. 1999). The name TWO originates from the two layers used in early versions of the model to describe an assumed abrupt change in the regolith thermo-physical properties with depth. Recently Vasavada et al. (2012) revised the model by allowing both thermal conductivity and bulk density to gradually increase with depth. The updated model accurately reproduced 513,738 Diviner temperature measurements along the lunar Equator taken over a period of 2.5 years and covering 4 complete diurnal lunar cycles (Vasavada et al. 2012). TWO uses 'first principles' to simulate point-level surface energy balance and a 1-D subsurface heat flow. The model calculates subterranean transport and storage of heat as a function of depth-varying thermal conductivity and bulk density of the lunar regolith.

We chose a validated physics-based model to verify Equations (10) and (11a) over the actual Diviner measurements, because the latter do not yet provide the temporal and global spatial coverage necessary for a robust estimation of the Moon mean annual equilibrium surface temperature. In addition, the Diviner data set literally contains millions of raw radiometric measurements that require a dedicated project to properly screen and convert into temperature readings. Since TWO has been shown to accurately reproduce Diviner-measured temperatures under a wide range of conditions and the Moon's regolith appears to be spatially highly homogeneous in terms its thermo-physical properties (Vasavada et al. 2012), we assumed that this model would yield sufficiently accurate results for all lunar locations.

Figure 5 depicts the average diurnal course of surface temperature at the lunar Equator simulated by the revised TWO models. As illustrated in Figure nine of Vasavada et al. (2012), this temperature curve agrees quite well with hundreds of thousands of Diviner measurements. The curve yields a mean equatorial temperature of 213 K (−60.15 C). In accordance with Hölder's inequality, the warmest latitude on the Moon is on average 57.1 K cooler than the lunar effective emission temperature (≈270 K) computed from Eq. (3).

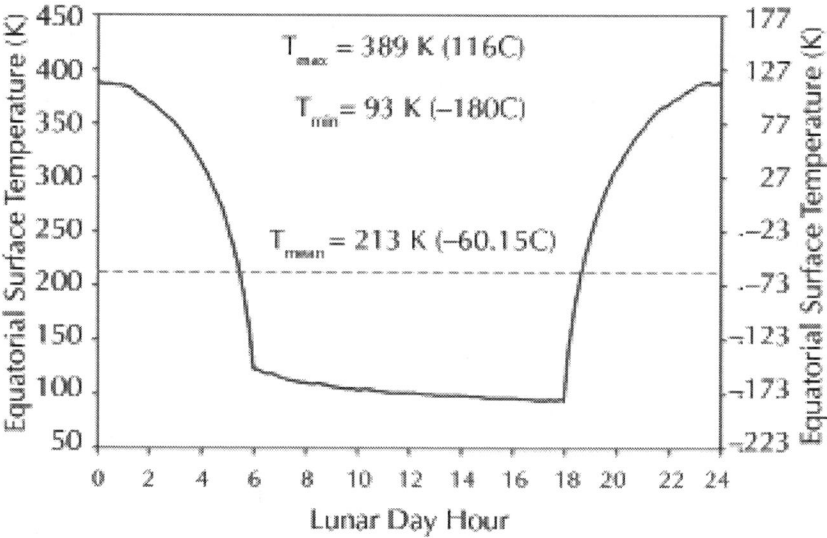

Figure 5: Typical diurnal course of the Moon's equatorial surface temperature according to Diviner radiometric measurements and simulations by the revised TWO models (based on data in Figure nine (a) of Vasavada et al. 2012). Also shown are the maximum (T_{max}), minimum (T_{min}) and mean (T_{mean}) temperature of the lunar Equator. In agreement with Hölder's inequality, $T_{mean} = 213$ K is about 57 K cooler than the Moon's effective emission temperature (270.2 K) calculated from Eq. (3).

We employed a three-step approach to obtain an independent estimate of the Moon's mean annual global surface temperature. First, we ran the TWO models as described by Vasavada et al. (2012) at every 5 degree latitude from the lunar Equator to the Poles calculating an annual-mean temperature for each latitude that represented both Northern and Southern lunar Hemisphere. The simulation employed a temporal resolution of 0.01 lunar hours and actual orbital characteristics of the Moon derived from publically available ephemerides of the Navigation and Ancillary Information Facility at the Jet Propulsion Laboratory. Solar irradiance was set to $S_o = 1,360.9$ W m^{-2} at a distance of 1 AU based on recent satellite observations reported by Kopp and Lean (2011). The irradiance was allowed to vary during the course of a year in accordance with Earth's small orbital eccentricity. The shortwave albedo of regolith was modeled as a function of the solar incidence angle using Eq. (6) with the normal albedo set to $A_o = 0.105$

based on Diviner observations (Vasavada et al. 2012). The thermal emissivity of regolith was assumed to be spatially invariant and equal to 0.98 (Vasavada et al. 2012). TWO was run for all 5-degree latitude bands over multiple years to allow equilibration of the annual-mean temperatures.

Next we fitted a 6th order polynomial to the modeled latitudinal temperature averages to derive a continuous function that smoothly describes the variation of the lunar annual-mean temperature T (K) with latitude L (rad):

$$
\begin{aligned}
T(L) = 212.9 &+ 9.919L - 119.814L^2 \\
&+ 307.116L^3 - 466.244L^4 \\
&+ 321.317L^5 - 84.973L^6
\end{aligned}
$$

$$(12)$$

Finally, the Moon's global mean temperature (T_{moon}) was calculated via integration of $T(L)$, i.e.

$$
T_{moon} = \int_0^{\pi/2} T(L) \cos L \, dL
$$

$$(13)$$

Where $\cos L$ is a polar coordinate area-weighting factor. Equation (13) yielded $T_{moon} = 197.3$ K. To our knowledge, this is the first physically robust estimate of the Moon's true average global surface temperature reported in the scientific literature. Figure 6 displays the results from the above computational approach. Note that the entire lunar latitudinal temperature curve lies well below the 270.1 K effective emission temperature derived from Eq. (3) with T_{moon} being nearly 73 K cooler than T_e. This illustrates the physical incompatibility between T_e and T_{moon}, which is mathematically explained by Hölder's inequality.

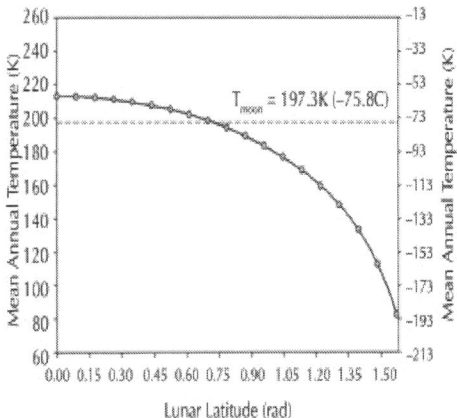

Figure 6: Mean annual temperature of the lunar surface as a function of latitude according to results from the validated TWO models (Vasavada et al. 1999, 2012) (black dots). The smooth curve represents a 6th-order polynomial (Eq. 12) fitted through the latitudinal temperature averages via a least-squares regression. The Moon mean annual global temperature, $T_{moon} = 197.3$ K (marked by a horizontal dashed line) was estimated through integration of the polynomial (12) using formula (13).

Verification Results

In order to properly verify the new model against the above independent estimate of the Moon mean surface temperature we used equivalent values for the driving variables in the formulas (10) and (11a) to those employed in the TWO thermo-physical model, i.e. $S_o = 1,360.9$ W m^{-2} and $\varepsilon = 0.98$. The effective shortwave albedo was set to $\alpha_e = 0.131$ according to Eq. (9). To obtain an estimate of the effective heat storage fraction (η_e) in Eq. (10) we analyzed output from the TWO models. First, we computed the annual fraction of the absorbed solar flux conducted into regolith (η) at several latitudes in order to evaluate its meridional variation. Latitudinal η values were calculated as ratios of the cumulative outgoing nighttime heat fluxes to the total daily-absorbed solar fluxes over the course of a typical lunar year. We found that η (L) increases non-linearly with latitude (Figure 7a). Such a functional relationship cannot be directly incorporated into equations (10) and (11a), since the integral formula calls for a *single* effective value η_e.

To estimate the latter we plotted $[1 - \eta(L)] \cos L$ versus $\cos L$ to discover a tight linear relationship between these variables with virtually zero intercept (Figure 7b).Since in the context of spherical integration, both $\cos L$ and $\cos \theta$ vary over the same numerical range $(0-1)$, the term $[1 - \eta(L)] \cos L$ is equivalent to $(1 - \eta) \cos \theta$ in the daytime portion of Eq. (5). Hence, one can use the slope s of the linear regression in Figure 7b to calculate an *effective*heat storage fraction η_e for Eq. (10). Specifically, the equality $[1 - \eta(L)] \cos L = s \cos L$ implies $(1 - \eta_e) = s$, which effectively neutralizes the meridional variation of the heat storage fraction $\eta(L)$ in the daytime portion of Eq. (5). Since $s = 0.99029$ (Figure 7b), we obtain $\eta_e = 1 - s = 0.00971$. Hence, the lunar surface effectively stores about 1% of the absorbed solar flux into the regolith as heat. This energy is subsequently released at night giving the dark side of the Moon a significantly higher surface temperature than expected from the cosmic background radiation alone.

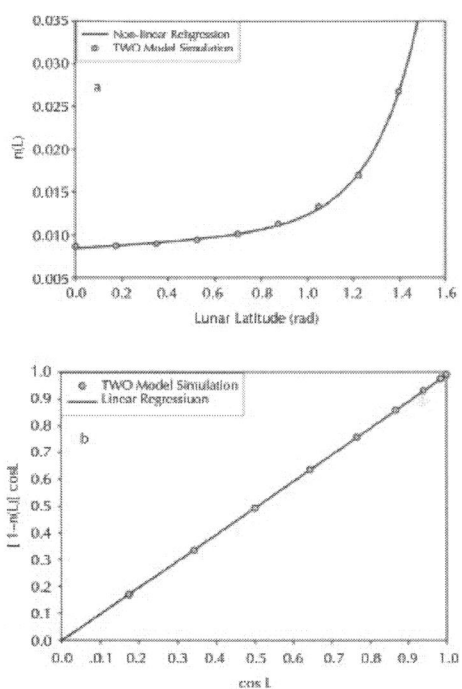

Figure 7: Relationship between the heat storage fraction of lunar regolith (η) and latitude (L): a) variation of η as a function of L according to the TWO

thermo-physical model output (Vasavada et al. 2012); b)emergent linear relationship between $[1-\eta(L)]\cos L$ and $\cos L$.

Using the above values of S_o, α_e, η_e and ε in either Eq. (10) or (11a) produces $T_{na} = 200.4$ K for the mean surface temperature of the Moon. This estimate is 3.1 K higher than $T_{moon} = 197.3$ K inferred from the TWO model (Eq. 13). However, considering the 72.8 K difference between T_e calculated from Eq. (3) and T_{moon}, it appears that equations (10) and (11a) provide a much more accurate estimate of the Moon average surface temperature compared to Eq. (3). Based on Hölder's inequality, we expect this to be the case for any ASCO. It is worth noting that Eq. (15) in Rubincam (2004) and Equation four in Leconte et al. (2013), which are similar to our Eq. (11a) without the heat storage term $\Phi(\eta_e)$, yield a 44.5 K lower global Moon temperature than T_{moon}. This demonstrates the critical importance of $\Phi(\eta_e)$. Without this enhancement factor our analytic model would have failed the verification against NASA Moon temperature data. Nevertheless, it is informative to explore the reasons for the small discrepancy between T_{na} predicted by Eq. (11a) and T_{moon}.

A close examination of Eq. (5) reveals that its nighttime portion contains the product $\eta \cos\theta$, which is numerically equivalent to $\eta \cos L$ in the context of spatial integration. If $\eta \cos L$ is plotted versus $\cos L$ using data from Figure 7a, a linear relationship emerges (Figure 8) similar to one in Figure 7b, but with an intercept that is significantly different from zero, i.e. $\eta \cos L = a\cos L + b$, where $a = 0.00458$ and $b = 0.00413$. This means that the term $\eta \cos\theta$ in the nighttime portion of Eq. (5) effectively linearizes the variation of η with latitude (Figure 7a) in the context of integral (7), but does not completely neutralize it as achieved by the term $(1-\eta)\cos\theta$ in the daytime portion of the same equation (Figure 7b). Hence, η_e derived from daytime conditions does not have the exact same mathematical meaning and magnitude in the nighttime portion of Eq. (10). Indeed, it can be shown that integrating the nighttime portion of Eq. (5) upon substituting $\eta \cos L$ with $a\cos L + b$ yields a complex solution. In other words, the slight overestimation of the Moon global surface temperature by equations (10) and (11a) appears to be the result of using an η_e value in the nighttime portion of the formula derived from a relationship that is strictly valid for daytime conditions (Figure 7b).

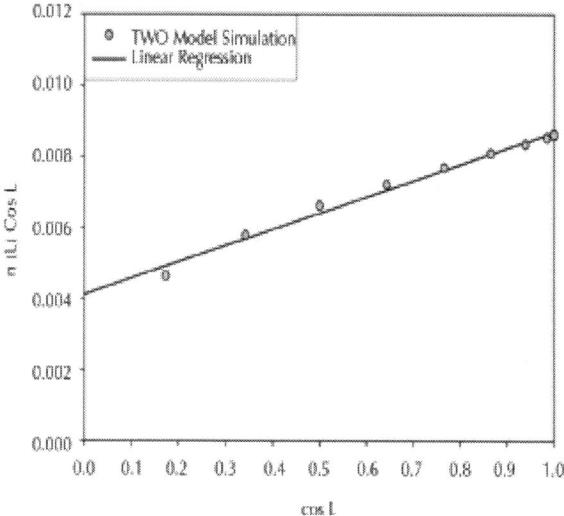

Figure 8: Emergent linear relationship between $\eta(L)\cos L$ and $\cos L$ based on modeled data in Figure 7a.

Refinement of the Analytic Temperature Formula for Airless Bodies

The above analysis suggests the need to adjust η_e in the nighttime portion of Eq. (10) in order for the analytic model to more accurately predict average global temperatures. Using data from Figure 8, we estimated a compensation factor 0.754 for η_e in the nighttime portion of Eq. (10). T model:

$$T_{na} = \frac{2}{5}\left\{ \frac{\left[(1-\eta_e)\,S_o\,(1-\alpha_e) + R_c + R_g\right]^{5/4} - (R_C + R_g)^{5/4}}{(1-\eta_e)\,S_o\,(1-\alpha_e)(\epsilon\sigma)^{1/4}} \right.$$
$$\left. + \frac{\left[0.754\,\eta_e\,S_o\,(1-\alpha_e) + R_c + R_g\right]^{5/4} - (R_C + R_g)^{5/4}}{0.754\,\eta_e\,S_o\,(1-\alpha_e)(\epsilon\sigma)^{1/4}} \right\}$$

$$(14)$$

Similar to Eq. (10), here one can also safely assume $R_c = 0.0$ if $S_o > 0.15$ W m^{-2} and $R_g = 0.0$ in most cases. This reduces Eq. (14) to (11a) with the regolith thermal enhancement factor redefined as:

$$\Phi(\eta_e) = (1 - \eta_e)^{0.25} + 0.932\, \eta_e^{0.25}$$

$$(15)$$

Where $0.932 = 0.754^{0.25}$. Thus, we now arrive at a simple yet robust and sufficiently accurate analytic expression for calculating the average surface temperatures of airless celestial bodies when $S_o > 0.15$ W m^{-2}:

$$T_{na} = \frac{2}{5} \left[\frac{S_o\,(1 - \alpha_e)}{\epsilon\sigma} \right]^{0.25} \left[(1 - \eta_e)^{0.25} + 0.932\, \eta_e^{0.25} \right]$$

$$(16)$$

In rare cases, where $S_o \le 0.15$ W m^{-2}, one must use formula (14) instead. Equation (14) is also recommended for airless bodies with a significant contribution of heat to the surface (R_g) by geothermal sources. Equation (16) yields 197.1 K for the Moon, a value within the uncertainty of the best estimate derived from the TWO thermophysical model.

We hypothesize that regolith-covered ASCOs would have similar effective albedos and heat storage coefficients because, in the absence of atmosphere, the pulverization of surface materials by micrometeoroids and cosmic radiation becomes the predominant geologic process creating a top substrate of similar particle-size distribution and thermophysical properties. Hence, it might be reasonable to employ the Moon-based value of $\eta_e = 0.00971$ in estimating the average surface temperatures of other airless bodies such as Mercury, for example. Using a solar irradiance $S_o = 9,086.7$ W m^{-2} (corresponding to Mercury's average distance of 0.387 AU to the Sun) and a plausible albedo range 0.068–0.142 (Mallama et al. 2002) in Eq. (16) yields $315.8 \le T_{na} \le 322.4$ K for that planet. Note that this estimate is ~120 K lower than the one derived from Eq. (3) (440 K) and currently quoted as Mercury's 'average temperature' (Williams 2014). Planetary science should soon be able to verify our prediction of Mercury's mean surface temperature using remote infrared measurements provided by the NASA MESSENGER robotic spacecraft.

RESULTS AND DISCUSSION

Aggregation Errors of the Effective Emission Temperature Formula

The above discussion reveals that effective emission temperatures calculated from Eq. (3) tend to be significantly higher than any long-term temperature averages on the surface of ASCOs. This discrepancy is mathematically explained by Hölder's inequality (4b). Having derived a new theoretically robust formula for the mean surface temperatures of ASCOs, we can now quantify the aggregation errors inherent in the standard formula (3).

We note that T_e and T_{na} have similar functional forms, i.e. both temperatures depend on the 4th root of solar irradiance. Hence, each formula can be written as $T = c S_o^{0.25}$, where c is a bulk coefficient combining surface albedo, infrared emissivity, regolith heat storage, and the SB constant. Denoting such bulk coefficients as c_e and c_{na} in equations (3) and (16) respectively, we obtain:

$$c_e = \left[\frac{1 - \alpha_e}{4 \, \epsilon \, \sigma} \right]^{0.25}$$

$$(17)$$

And

$$c_{na} = \frac{2}{5} \left[\frac{1 - \alpha_e}{\epsilon \, \sigma} \right]^{0.25} \left[(1 - \eta_e)^{0.25} + 0.932 \, \eta_e^{0.25} \right]$$

$$(18)$$

This consolidation of variables allows us to define two error functions, E_r (%) and E_a (K) quantifying the relative and absolute errors of Eq. (3) respectively, i.e.

$$E_r = \frac{T_e - T_{na}}{T_{na}} 100 = \frac{c_e - c_{na}}{c_{na}} 100$$

(19)

$$E_a = T_e - T_{na} = (c_e - c_{na}) S_0^{0.25}$$

(20)

In order to evaluate E_r and E_a we employ radiative and thermo-physical data for the Moon. Note that the particular choice of parameter values is not important as long as both formulas (17) and (18) use the same ones. Upon substituting $a_e = 0.13$, $\varepsilon = 0.98$, and $\eta_e = 0.00971$ in equations (17) through (20), we obtain $c_e = 44.48$, $c_{na} = 32.45$, $E_r = 37.1\%$ and $E_a = 12.03 S_0^{0.25}$. A greater heat storage fraction resulting from a higher thermal conductivity of the regolith would produce a larger c_{na}, thus boosting T_{na} towards T_e. However, based on the most likely upper limit of $\eta_e \approx 0.02$ for ASCOs, we estimate the minimum errors of the emission-temperature formula to be in the order of $E_r = 31.4\%$ and $E_a = 10.6 S_0^{0.25}$.

The above analysis reveals that Eq. (3) overestimates the average surface temperatures of ASCOs by about 37% with the absolute error of T_e increasing proportionally to the 4th root of solar irradiance (Eq. 20). Hence, T_e and T_{na} are not physically comparable. The numerical bias of T_e becomes particularly evident when comparing ASCOs orbiting at different distances from the Sun. For example, according to Eq. (3), Mercury (at 0.387 AU) should be about 172 K warmer than the Moon (at 1 AU). However, the theoretically correct formula (16) indicates an average temperature difference of no more than 125 K between the two bodies. Hence, Eq. (3) is not suitable for comparing the thermal regimes of planets as suggested by Leconte et al. (2013). Viewed across a range of planetary environments, the effective emission temperature shows no meaningful relationship to actual surface temperatures of bodies either with or without an atmosphere. In view of these results, we propose equations (14) and (16) as a new analytic standard for predicting the mean surface temperatures of airless spherical bodies.

Moon as a Natural Airless Equivalent of Earth

Our approach to evaluating Earth's total ATE rests on the supposition that, without an atmosphere, our Planet would be on average as cold as the Moon. Verifying this assumption requires addressing the following questions: 1) would the Earth surface in the absence of atmosphere have the same radiative and thermo-physical properties as the lunar regolith? And 2) is an airless Earth thermally equivalent to a hypothetical Earth with an atmosphere devoid of greenhouse gases? In other words, is ATE fully explainable by the radiative effect of heat-absorbing gases? Equations (14) and (16) provide a suitable framework for investigation. There are 4 variables in Eq. (16) impacting the average surface temperature of ASCOs: solar irradiance (S_o), shortwave albedo (α_e), surface long-wave emissivity (ε), and the regolith's effective heat storage coefficient (η_e).

Since Moon and Earth orbit the Sun at the same distance, they receive equal amounts of solar radiation and have the same S_o. Serendipitously, the Moon effective albedo $\alpha_e = 0.131$ nearly equals Earth's present surface average cloudless albedo (0.122–0.13) inferred from satellite- and ground-based observations (Stephens et al. 2012; Wild et al. 2013). This is in spite of the fact that our Planet has highly reflective regions such as deserts, glaciers, and Polar Ice Caps that are absent on the Moon. However, the high reflectivity of these Earth surfaces is counterbalanced by the low albedo of the World's Oceans. Aside from this coincidental similarity of surface albedos between present-day Earth and the Moon, one can also argue that, in the absence of atmosphere, Earth would have no liquid oceans and/or exposed glaciers, since these require an atmospheric pressure (P) and temperature (T) above the triple point of water to exist, i.e. $T > 273.2$ K and $P > 611.73$ P_a (Cengel and Turner 2004). Without an atmosphere, the surface of our planet would be subjected to the same geologic processes that presently govern regolith formation on the Moon (e.g. bombardment by cosmic radiation and micrometeorites). Hence, an airless Earth would likely have a surface soil layer of similar radiative and optical properties (shortwave albedo and long-wave emissivity) as the lunar regolith. The uncertainty of the ATE estimate associated with Earth's airless albedo is further discussed below.

The effective heat storage fraction η_e is the only term among the independent variables in equations (14) and (16) that significantly differs between the Moon and present-day Earth. While lunar regolith stores on average less than 1% of the absorbed solar flux in the ground as heat, landmasses on Earth typically conduct 5%–6% of the daytime absorbed short- and long-wave radiation into the subsurface known as *soil heat flux*. Oceans store an even greater fraction due to the high thermal conductivity and volumetric heat capacity of water. On average, Earth conservatively stores between 7% and 8% of the total daytime absorbed radiation into the subsurface to be released as heat at night. Assuming $\eta_e = 0.075$ as an average value in Eq. (15) produces a significantly larger soil thermal enhancement factor $(0.075) = 1.47$ for Earth compared to Φ $(0.00971) = 1.29$ for the Moon. This raises two additional questions: a) what enables the Earth surface to store substantially more heat than the lunar regolith? And b) does a higher surface heat storage (i.e. a larger η_e) make Earth a poor comparison to the Moon for the purpose of ATE evaluations? To answer these we must analyze the factors controlling η_e.

Effect of Surface Thermal Conductivity on Regolith Heat Storage

At any given latitude, the heat-storage fraction η depends on the cumulative ground heat flux G_o (W m^{-2}) absorbed during the course of a typical day. Instantaneous ground heat fluxes, in turn, are functions of the substrate apparent thermal conductivity k (W m^{-1} K^{-1}) and the time-varying vertical temperature gradient ($\partial T (t)/\partial z$, K m^{-1}) at the surface (e.g. Campbell 1985). Thus, G_o can be described at any latitude as

$$G_o = \left[\int_{t1}^{t2} \left(k \frac{\partial T(t)}{\partial z} \right) dt \right] / L_d$$

(21)

Where t_1 and t_2 are the times of sunrise and sunset, respectively, and $L_d = t_2 - t_1$ is the day length at that latitude.

In the dynamics of heat flow, $\partial T (t)/\partial z$ varies throughout the day for a particular k as a function of the changing solar forcing at the surface. For a given radiation intensity, however, the temperature gradient tends to be inversely related to k, i.e. a higher thermal conductivity tends to produce smaller vertical temperature gradients, while a lower k results in larger $\partial T(t)/\partial z$. This curtails the sensitivity of G_o to changes in k making G_o vary nonlinearly with conductivity albeit in the same direction. Thus, it takes a relatively large increase in k to produce a moderate rise in G_o for otherwise equal conditions. For example, results from a sensitivity test of the TWO model reported by Vasavada et al. (2012, their Figure six (a)) indicate that a 30-fold boost of the regolith thermal conductivity only causes a 5-fold increase of the emitted nighttime heat flux. Nevertheless, k controls the size of both G_o and η. To understand the factors controlling the magnitude of k it is informative to compare empirical models of thermal conductivity for the lunar regolith and Earth's soils based on in-situ measurements. According to Vasavada et al. (2012), the thermal conductivity of the lunar surface varies with depth (z, m) and temperature (T, K) as

$$k(z, T) = k_d - (k_d - k_s) \exp(-z/0.06)$$
$$+ 2.7k_s \left(\frac{T}{350} \right)^3$$

$$(22)$$

Where $k_s = 0.0006$ and $k_d = 0.007$ (W m^{-1} K^{-1}) are respective conductivities at the surface and at ~0.1 m depth. The factor 2.7 is a ratio of the radiative to solid component of k at $T = 350$ K. The apparent thermal conductivity of a mineral soil on Earth λ (w), which includes the effects of sensible and latent heat transport, can be described by the following function (Cass et al. 1984; Campbell 1985):

$$\lambda(w) = A + Bw(1 - f_r) - (A - D) \exp\{-[Cw(1 - f_r)]^4\}$$

$$(23)$$

Where w is the volumetric soil moisture content (m^3 m^{-3}), fr is the volumetric fraction of rocks (i.e. particles with a diameter greater than

2 mm), and A, B, C and D are empirical coefficients depending on f_r, total soil porosity (p_s, m^3 m^{-3}), and percent clay (C_l) in soil, so that

$$A = \frac{1.57 - p_s(1-f_r)}{1 - 0.52[1 - p_s(1-f_r)]} - 2.8p_s[1 - p_s(1-f_r)]$$
$$B = 2.8[1 - p_s(1-f_r)]$$
$$C = 1 + 26C_l^{-0.5}$$
$$D = 0.03 + 0.7[1 - p_s(1-f_r)]^2$$

According to Eq. (22), at 0.5 cm depth and typical Earth temperatures (i.e. 263 K–310 K), the thermal conductivity of the lunar regolith is only $0.0018 \leq k \leq 0.0022$ W m^{-1} K^{-1}. For comparison, a completely dry soil of similar texture, bulk density, and rock content (i.e. $p_s = 0.494$ m^3 m^{-3}, $f_r = 0.028$, and $C_l = 0.01$) on Earth has a thermal conductivity $\lambda(0) = 0.219$ W m^{-1} K^{-1} according to Eq. (23). Increasing the soil moisture content to its maximum ($w = p_s$) boosts that conductivity to λ (0.494) = 1.473 W m^{-1} K^{-1}. Hence, a substrate of similar particle size distribution and bulk density as the lunar regolith is over 100 times more conductive to heat on Earth than it is on the Moon. A moisture-saturated soil of the same type on Earth has over 700 times greater thermal conductivity than the lunar regolith.

The immense difference in the ability to conduct heat between Moon and Earth can be explained by analyzing the components of the apparent thermal conductivity, i.e. solid, radiative, and convective (the latter component includes sensible and latent heat transport). Solid conduction results from the vibrational transfer of energy between atoms comprising the material lattice of regolith particles. This type of conduction increases with particle size and bulk density of the substrate. The radiative component arises from radiant heat exchange between regolith grains and is proportional to the third power of the grains' absolute temperature (Vasavada et al. 2012). At relatively low bulk densities and high temperatures found near the surface, radiant heat exchange typically dominates the regolith thermal conductivity in airless environments. The convective component of heat conduction is due to collision of gas molecules residing in the space between soil particles and requires the presence of an atmosphere to operate.

Since sensible and latent heat fluxes are several orders of magnitude more effective in transporting energy compared to radiation or solid conduction, the interstitial micro-convection becomes the predominant mechanism of thermal conduction in porous media immersed in an atmosphere. Indeed, laboratory experiments by Presley and Christensen (1997) have shown that the apparent thermal conductivity of dry regolith increases with the 2/3-power of atmospheric pressure between 0 P_a and 1,000 P_a. These results have recently been confirmed by in-situ measurements of Martian soil made by the Thermal and Electrical Conductivity Probe (TECP) aboard the Phoenix Lander (Zent et al. 2010). The observed surface thermal conductivity in the northern polar region of the Red Planet (~0.085 W m^{-1} K^{-1}) is consistent with measurements made by Presley and Christensen (1997) in a simulated Martian atmosphere on Earth. In other words, thanks to the presence of a tangible atmosphere, Mars has a nearly 50-time higher thermal conductivity than the Moon. Hence, it is the presence of an effective vacuum and the related lack of gaseous micro-convection within the lunar regolith that makes the Moon such a poor heat conductor. This implies that regolith-covered ASCOs can be expected to have similarly low surface thermal conductivities. The current Earth surface is vastly more conductive to heat than either lunar regolith or Martian soil because of the sizable atmospheric pressure present on our planet. Earth's thermal conductivity is further boosted by moisture (liquid water), which cannot exist without ATE. In the absence of atmosphere, there would be no interstitial convection to boost η_e. Therefore, an airless Earth would have a surface of similar thermo-physical properties as the present lunar regolith. The strong dependence of surface thermal conductivity and η_e on atmospheric pressure and soil moisture lends additional physical support to the notion that Earth's overall ATE ought to be evaluated with respect to an equivalent *airless* environment rather than a hypothetical atmosphere devoid of greenhouse gases.

Effect of Planet's Rotation Rate on Regolith Heat Storage

Propositions have been made in the literature that a planet spin rate (ω, Hz) can affect the average surface temperature of ASCOs. Specifically, it has been suggested that a higher ω would cause a planet's T_{na} to approach T_e (e.g. Smith 2008). A comprehensive mathematical analysis of the effect of rotation on surface temperature is beyond the scope of this study. Here we shall only briefly explore the expected equilibrium response of a planet's global surface temperature to a sustained change in spin rate according to the standard theory of heat flow.

The effective heat storage fraction η_e is the only variable in Eq. (16) that might potentially be affected by a change in planet's spin rate. Since η_e is a function of $\eta(L)$ (Figure 7), first we need to investigate whether rotational speed can influence the equilibrium heat-storage coefficient ηL at any latitude L. We begin with the mathematical definition of ηL as a ratio of the cumulative daytime ground heat flux to the daily total absorbed solar flux adopted in this study, i.e.

$$
\eta_L = \left\{ \int_{t1}^{t2} \left(k \frac{\partial T(t)}{\partial z} \right) dt \right\} \bigg/ \left\{ \int_{t1}^{t2} S_L [1 - A(\theta, t)] \cos[\theta(t)] dt \right\}
\tag{24}
$$

where S_L is the maximum incident solar radiation at latitude L, $\theta(t)$ is the solar zenith angle as a function of time t, $A(\theta, t)$ is the surface albedo as a function of $\theta(t)$, and $(t_2 - t_1)$ is the day length. This definition assumes that, in equilibrium, all the energy conducted into regolith during daytime is completely released at night warming the surface on the dark side of the planet. Obviously, an increase of rotational speed would shorten the day length, which will reduce both the amount of absorbed daily shortwave radiation and the period of heat conduction into the regolith. However, the key question is how would a change in the spin rate affect the numerator of Eq. (24) via instantaneous heat fluxes? To answer it we analyze an idealized analytic solution to the heat flow equation discussed by Campbell (1985) and Ochsner et al. (2007).

According to the classic theory of heat flow in porous media, increasing ω reduces the depth to which the surface heat wave can propagate. This is mathematically described by the so-called 'damping depth' (Z_d, m) as referred to in soil science (Campbell 1985) or 'thermal skin depth' as named in planetary science (Leyrat et al. 2011):

$$Z_d = \sqrt{\frac{k}{C_v\,\omega}}$$

(25)

Here, Cv is the volumetric heat capacity (J m−3 K−1) of regolith, which is a product of specific heat capacity cp (J kg−1 K−1) and bulk density ρ (kg m-3), i.e. Cv = ρ cp. Physically, Zd is the depth, where the diurnal amplitude of the surface heat wave is reduced by a factor of e (i.e. 2.718 times). Another important physical property of the regolith is the thermal inertia defined as $I_T = \sqrt{kC_v}$ $(jm^{-2}k^{-1}s^{-1/2})$. It measures the 'resistance' of the substrate's temperature to change. The higher the thermal inertia the more energy the regolith must absorb in order to increase its temperature by 1 K. The damping depth can also be expressed in terms of thermal inertia as

$$Z_d = \frac{I_T}{C_v\sqrt{\omega}}$$

(26)

Since both c_p and k of the lunar regolith increase with temperature at about the same rate (Hemingway et al. 1973; Vasavada et al. 2012), Z_d has almost no sensitivity to temperature. However, since thermal conductivity strongly depends on air pressure (Presley and Christensen 1997), Z_d varies broadly between Earth, Mars and the Moon primarily as a function of atmospheric pressure. The sensitivity of Z_d to variations in ω increases with air pressure and the thermal conductivity of regolith.

The diurnal amplitude of the heat-flux wave (M_s) decreases exponentially with regolith depth and planet's spin rate according to the formula (Ochsner et al. 2007):

$$M(z, \omega) = M_s \exp(-z/Z_d)$$
$$= M_s \exp\left(-z\sqrt{C_v\,\omega/k}\right)$$

$$(27)$$

Where $M(z, \omega)$ is the amplitude at depth z and rotational frequency ω, and M_s is the heat-wave amplitude at the surface. The relative rate of change in $M(z, \omega)$ with ω is given by the first partial derivative of Eq. (27):

$$\frac{\partial M_z}{\partial \omega} = -0.5\,z\,\sqrt{C_v/(\omega k)}\,\exp\left(-z\sqrt{C_v\,\omega/k}\right)$$

$$(28)$$

Where $M_z = M(z, \omega)/M_s$. Equations (27) and (28) suggest that an increasing spin rate has a greater impact of reducing the subterranean diurnal temperature amplitude near the surface than it does at depth. This implies that the *annual mean* temperature of the subsurface is not impacted by ω. A planet spin rate also affects the time lag between heat flux maxima at the surface and at depth z. This lag decreases in proportion to $1/\omega\sqrt{}$ (Ochsner et al. 2007), which means that a faster rate of rotation brings the subsurface heat wave closer in phase to the surface heat wave.

Equation (25) through (28) collectively suggest that a faster rotational speed causes less heat to be conducted into the ground under steady-state conditions. This is due to a reduction of the vertical temperature gradient at the surface, which is a key factor controlling the magnitude of instantaneous ground heat fluxes (Eq. 21). The vertical gradient $\partial T/\partial z$ reaches maximum for a particular illumination when the incident solar flux changes slow enough to stay in equilibrium with the outgoing infrared flux. Increasing the rotational speed causes incident solar radiation to vary faster than the rate of $\partial T/\partial z$ formation set by the regolith thermal inertia. As a result, the temperature gradient begins to deviate from its potential strength. The faster the rotation, the greater the deviation of $\partial T/\partial z$ would be for a given ground thermal inertia.

Thus, increasing a planet's angular velocity *decreases* the average ground heat flux via reduction of instantaneous heat fluxes. This causes the ratio in Eq. (24), i.e. the solar flux fraction stored into the ground to remain conservative across planet spin rates. The Law of Energy Conservation dictates that a change in rotational speed may only affect the magnitude of the diurnal temperature amplitude at the surface but not the diurnal mean, i.e. rotation solely acts to redistribute the total available energy between daytime and nighttime hemispheres through the planet's thermal inertia. As the rotation frequency increases beyond a certain threshold, the surface temperature amplitude begins to shrink, thus flattening the diurnal heat wave without affecting the diurnal mean. This frequency threshold depends on the absolute magnitude of the surface thermal inertia, i.e. the greater the inertia the slower the rotational speed, at which the diurnal temperature amplitude becomes affected. Since regolith-covered ASCOs such as the Moon are expected to have a very small thermal inertia due to a low thermal conductivity of the regolith in vacuum, it takes a rather fast axial rotation to noticeably impact the diurnal temperature amplitude at the surface. This analysis suggests that ω cannot affect η_e and the average surface temperature of a planet. The heat storage fraction can only be altered by a significant change in the apparent thermal conductivity of the regolith, which requires the introduction of a qualitatively different environment such as adding atmospheric pressure to the surface.

Magnitude and Components of Earth's Atmospheric Thermal Effect

The above discussion leads to the conclusion that Earth's *total* ATE must be evaluated with respect to Earth's hypothetical *airless* self. The thermal environment of the Moon offers a perfect natural airless equivalent of our Planet. Therefore, using either T_{na} = 197.1 K calculated from Eq. (16) or T_{moon} = 197.3 K derived from the TWO thermo-physical model, we obtain:

$$\text{ATE} = T_s - T_{na} = 287.6 - 197.1 = 90.5 \text{ K}$$

And

$$\text{ATE} = T_s - T_{moon} = 287.6 - 197.3 = 90.3 \text{ K}$$

Accepting Eq. (16) as the proper analytic model for calculating the average surface temperature of airless spherical bodies allows us to produce an alternative estimate of T_{na} using Earth's present surface albedo of 0.122 inferred from satellite observations (Stephens et al. 2012). The result is $T_{na} = 197.6$ K, which translates into ATE = 287.6–197.6 = 90.0 K. This is only 0.3K–0.5 K lower than the above ATE estimates based on the Moon albedo of 0.13. In order to claim robustness of our ATE estimate, however, we must evaluate the uncertainty of T_{na} associated with a plausible range of Earth albedos in the absence of atmosphere. There is currently no rigorous quantitative method available for calculating the albedo of a hypothetical airless Earth; hence, one must use physical reasoning to obtain a proper value range. Two conditions need be met when utilizing observational data from airless bodies in the Solar System to make a logical inference about Earth's albedo without an atmosphere:

1) One must only consider the regolith albedos of *ice-free* airless bodies. This is so, because the solar heating at Earth's orbit is strong enough to quickly evaporate any exposed water ice on an airless surface. Ices of other gases such as CO_2, CH_4 and nitrogen also cannot form under a solar irradiance of 1,361 W m^{-2}. It is for this reason that significant amounts of water ice are only found on the airless Moon in permanently shadowed craters near the lunar poles (Colaprete et al. 2010; Spudis et al.2010). Hence, this condition excludes from consideration the high albedos of airless icy bodies such as Saturn's satellites Rhea, Dione, Tethys, Mimas, and Enceladus as well as Jupiter's moon Europa;

2) One should consider bolometric Bond albedos rather than geometric albedos in global energy-budget calculations. This is because Bond albedos are spherical, while geometric albedos are directional. Studies oftentimes only report geometric albedos of celestial objects, since these are directly measurable, while Bond albedos must be calculated and require knowledge of the hemispheric phase integral.

Airless bodies usually have Bond albedos that are lower than their geometric albedos.

Available surface reflectance data from the Solar System suggest that bolometric Bond albedos of ice-free regolith-covered airless bodies typically range from 0.068 to 0.16 (Mallama et al. 2002; Shestopalov and Golubeva 2011). Some small-sized asteroids mostly composed of rocks have lower Bond albedos than 0.068, but such values are not typical for larger regolith-covered bodies. Employing the above albedo limits with Eq. (16) yields a global average temperature for a hypothetical airless Earth 195.4 K $\leq T_{na} \leq$ 200.6 K. This translates into an ATE between 87.0 K and 92.2 K. Thus, one can formally quote Earth's ATE as 89.6±2.6 K, although we consider ATE = 90.5 K to be our best estimate. Therefore, the thermal effect of our atmosphere is 2.7 to 5 times stronger than currently assumed based on Eq. (3). According to our analysis, Earth's ATE varies spatially from 86 K at the Equator to about 148 K at the Poles.

In order to assess the contribution of Earth's present surface heat storage to the planet's global temperature, we set $\eta_e = 0.075$ and $\alpha_e = 0.294$ in Eq. (16). The result is a new reference temperature $T_{na2} = 213.0$ K that is 15.7 K higher than the Moon's present airless temperature (our true reference). This thermal enhancement is caused by a 7.7-fold increase of η_e above the corresponding lunar value. The larger heat storage fraction reflects the presence of an atmosphere and is a consequence of a much higher surface thermal conductivity on Earth due to air pressure. Note that increasing the albedo from 0.131 to 0.294 only partially offsets the enhancement effect of a larger η_e on global temperature. Thus, the daytime storage of heat by landmasses and oceans on Earth significantly contributes to our planet's ATE by raising the average nighttime temperatures. This implies that Earth's atmospheric effect has a sizable thermodynamic component that is independent of the greenhouse infrared back radiation. In other words, ATE includes more than just the radiative effect of greenhouse gases (GE), i.e. ATE = GE + TE, where TE is a Temperature Enhancement caused by thermodynamic (pressure-controlled) processes. The thermal effect of radiatively active gases is then obtained as a residual, i.e. GE = ATE − TE.

We must make an important clarification regarding the above decomposition of ATE. The subterranean heat storage (considered in our model) boosts the global surface temperature by effectively 'transporting' a fraction of the daytime absorbed solar flux to the night side of the planet via axial rotation. In the presence of atmosphere, however, the air- and oceanic currents (not considered in our airless model) will foster additional lateral transfer of heat that further increases the planet's average temperature. This energy transport via fluid motion includes advective and radiative components that are difficult to separate through a simple analysis. Hence, the thermodynamic portion of Earth's ATE is likely greater than 15.7 K in reality. However, an accurate assessment of the TE magnitude is a non-trivial task and entails the use of coupled atmosphere–ocean global circulation models, which is beyond the scope of this study. Therefore, the above TE estimate (~16 K) should merely be viewed as an indication that Earth's ATE has indeed a sizable thermodynamic component requiring further investigation. Indirect support for the existence of TE is also provided by the fact that the observed 158 W m^{-2} global atmospheric absorption of outgoing long-wave radiation (Stephens et al. 2012; Wild et al. 2013) cannot fully explain the hereto deduced ~90 K total ATE. The TE component inferred from our model analysis offers a new premise to the Greenhouse theory, which currently attributes 100% of Earth's ATE to an infrared heat trapping by greenhouse gases (e.g. Hansen et al. 1981; Peixoto and Oort 1992; Schmidt et al. 2010; Lacis et al. 2013).

We surmise that the radiative portion of ATE controlled by greenhouse gases might be larger than 33 K in reality. Such a conjecture is supported by a recent simulation study of Russell et al. (2013), which found using a streamlined 3-D coupled atmosphere–ocean model that Earth's global temperature would drop 44.4 C (to −30 C) if the atmospheric CO_2 concentration were reduced to 1/8 of its 1950 level. However, these authors did not examine the change in modeled global temperature under an atmosphere completely devoid of greenhouse gases. Had they done so, the outcome would probably have been a greater surface cooling than 44 C. These simulation results (if correct) when combined with our findings indicate that the long-term impact of increasing greenhouse-gas emissions on climate might be stronger than currently projected. Indeed, recent paleoclimate studies comparing proxy-derived temperatures and CO_2 concentrations for the mid-Pliocene to simulations by fully coupled models (Lunt et al. 2010; Haywood et al. 2013) found that the Earth System Sensitivity

(ESS) (defined as the equilibrium global surface temperature response to a sustained doubling of atmospheric CO_2 concentration including *all* feedbacks) might be 1.5 times higher than the conventional climate sensitivity to CO_2 and the water-vapor feedback simulated by climate models. Paleoclimate studies have also suggested that ESS might depend on the state of the climate system (e.g. Caballero and Huber 2013).

CONCLUSIONS

The observed global energy balance of celestial bodies is often used in conjunction with a simple form of the SB radiation law (Eq. 3) to calculate equilibrium effective radiating temperatures (T_e) that find a broad application in today's climate and planetary sciences. For more than 35 years, these calculated temperatures have been utilized to compare the thermal regimes of airless bodies, to quantify the strength of atmospheric greenhouse effects, and to assess the potential habitability of extrasolar planets. Thus, Earth's effective radiating temperature of ~255 K, which includes the albedo effect of clouds, is the basis for the popular 33 K estimate of the total background atmospheric warming a.k.a. Natural Greenhouse Effect. Although T_e is derived from a well-known physical law, a close examination of the meaning of this temperature in the context of planetary energetics and spherical geometry reveals two critical caveats. First, a global emission temperature computed from Eq. (3) using the planet's actual Bond albedo that includes the radiative effects of greenhouse gases and air molecules, cannot serve as a proper reference in quantifying the thermal effect of a planetary atmosphere. This is because a reference temperature is expected in this case to describe a thermal state in the *absence* of greenhouse gases or an atmosphere. However, Earth's 33 K GE estimate is based on a T_e value that violates such a condition. Zeng (2010) recognized this and proposed using an average land-ocean*surface* albedo in Eq. (3) instead of Earth's Bond albedo as a solution. He essentially argued that the appropriate reference temperature for calculating Earth's ATE is that of an airless Earth, which we concur with. Employing a 0.14 surface albedo in Eq. (3) Zeng arrived at GE≈20 K. However, this author did not address another more fundamental problem of Eq. (3) related to Hölder's inequality between integrals. The non-linearity of the SB radiation law coupled with a strong latitudinal variation of the absorbed solar flux across the surface of a sphere creates a mathematical condition that precludes *in principle* a correct calculation of the true global surface temperature from a spatially integrated radiative flux. In other words, due to Hölder's inequality, one always finds $T_e \gg T_{na}$. Leconte et al. (2013) acknowledged this phenomenon, but the actual magnitude of the inequality and its theoretical implications have not been fully analyzed prior to our study. We showed that the actual mean surface temperature of the Moon (197.3 K) is about 73 K cooler than the Moon's effective radiating temperature $T_e \approx 270$ K computed from Eq. (3) using the same albedo. This large discrepancy is due to the fact that Eq. (3) essentially yields a disk-average temperature instead of a spherical temperature mean. Most studies treat globally averaged radiative fluxes and their corresponding effective radiating temperatures as physically interchangeable quantities (e.g. Lacis et al. 2013). However, according to Hölder's inequality (4b), this is conceptually incorrect. While the average outgoing LW flux of a planet is an observable physical parameter, T_e derived from it using Eq. (3) is a mathematical abstraction with no physical analogue at, below or above the surface. Thus, planetary effective emission temperatures are not compatible with measured physical temperatures regardless of the albedo they are based on. In other words, T_e is a non-physical quantity with respect to a sphere. Consequently, comparing Earth's observed global mean surface air temperature (287.6 K) to any T_e is bound to produce numerically and theoretically misleading results. The conceptual distinction between T_e on one hand and T_{na} and T_s on the other arises from the mathematical understanding that mean planetary temperatures cannot in principle be inferred from globally averaged radiative fluxes. In this regard, our analysis demonstrates that evaluating the strength of a planet's ATE strictly requires the use of physical surface temperatures.

To properly account for Hölder's inequality, we derived a new expression for the mean surface temperature of airless bodies (Eq. 10 and 11a) based on analytic integration of the SB law over a sphere with explicit consideration of the effects of regolith heat storage and cosmic background radiation on nighttime temperatures. The new model was successfully verified against independent temperature data for the Moon provided by the Diviner Lunar Radiometer Experiment. Verification results suggested a small adjustment to the nighttime integration. The final equations (14) and (16) provide a substantially improved method for quantifying the average temperatures of airless bodies compared to the current SB formula (3). An error analysis employing the new model revealed that Eq. (3) overestimates the global physical temperature of airless bodies by about 37% with the absolute error increasing proportionally to the 4th root of the TOA stellar irradiance. This large bias follows from functional differences between equations (3) and (16). Based on these results, we propose equations (14) and (16) as a new analytic standard for calculating the average surface temperatures of airless bodies.

We presented evidence that the Moon is a perfect airless grey-body equivalent of Earth. A key element of this evidence is that the regolith heat storage fraction η_e, which has a critical impact on the global temperature (Eq. 16), strongly depends on atmospheric pressure through the surface thermal conductivity. We showed that air pressure significantly boosts the heat storage capacity of Earth compared to the lunar environment and significantly contributes to the overall thermal effect of our atmosphere. The presence of such a large thermodynamic component (TE) implies that, when it comes to assessing the total magnitude of ATE, an Earth *with* an atmosphere devoid of greenhouse gases is not physically equivalent to an Earth *without* an atmosphere. Hence, the *overall* thermal effect of a planetary atmosphere should be evaluated with respect to the mean surface temperature of an equivalent *airless* body calculated from Eq. (14) or (16). Combining Earth's observed global surface temperature with results from the new analytic model reveals that the total thermal effect of our atmosphere is about 90 K or 2.7 to 5 times stronger than currently assumed. At least 17% (15.7 K) of this ATE is due to thermodynamic factors that are independent of the atmospheric infrared back radiation. The non-radiative portion of Earth's ATE is likely greater than 15.7 K in reality due to horizontal heat transports by oceanic and atmospheric currents not considered in our model. The hereto identified thermodynamic component of ATE creates a new premise for the Greenhouse theory, which currently attributes 100% of the background atmospheric warming to a long-wave radiation trapping by greenhouse gases. Finally, our analysis suggests that the exact contribution of heat-absorbing gases to Earth's atmospheric effect will remain unknown until the non-radiative component of ATE is fully quantified. Therefore, further fundamental research is needed in atmospheric radiative transfer and

3-D tropospheric thermodynamics to better constrain the functional elements of Earth's atmospheric thermal effect.

AUTHORS' CONTRIBUTIONS

DV designed the overall study, derived the analytic model for estimating the average surface temperature of airless spherical bodies, performed the model verification, and wrote most of the article narrative. LR calculated the annual-mean surface temperature of the Moon and the heat storage coefficient of the lunar regolith using simulation results from the TWO thermo-physical model provided by Dr. Ashwin R. Vasavada at the Jet Propulsion Laboratory of California Institute of Technology. LR also contributed to the preparation of key sections of the manuscript. Both authors read and approved the final manuscript.

ACKNOWLEDGEMENTS

The authors thank Dr. Ashwin R. Vasavada at the Jet Propulsion Laboratory of California Institute of Technology for providing lunar equatorial temperature data and simulation results for the Moon using the TWO thermo-physical model as well as for insightful discussions regarding properties of the lunar regolith.

REFERENCES

1.   Abualrub NS, Sulaiman WT (2009) A note on Hölder's inequality. Int Math Forum 4(40):1993-1995

2.   Bandfield JL, Ghent RR, Vasavada AR, Paige DA, Lawrence SJ, Robinson MS (2011) lunar surface rock abundance and regolith fines temperatures derived from LRO diviner radiometer data. J Geophys Res 116:E00H02 doi:10.1029/2011JE003866

3.   Beckenbach EF, Bellman R (1983) Inequalities. Berlin: Springer Verlag.

4.   Blanco VM, McCuskey SW (1961) Basic physics of the solar system. Reading MA: Addison-Wesley.

5. Caballero R, Huber M (2013) State-dependent climate sensitivity in past warm climates and its implications for future climate projections. Proc Natl Acad Sci 110:14162-14167 doi:10.1073/pnas.1303365110

6. Campbell GS (1985) Soil physics with basic: transport models for soil-plant systems. Amsterdam: Elsevier Scientific Publications.

7. Cass BA, Campbell GS, Jones TL (1984) Enhancement of thermal water vapor diffusion in soil. Soil Sci Soc Am J 48:25-32

8. Cengel YA, Turner RH (2004) Fundamentals of thermal-fluid sciences. Boston: McGraw-Hill.

9. Colaprete A, Schultz P, Heldmann J, Wooden D, Shirley M, Ennico K, Hermalyn B, Marshall W, Ricco A, Elphic RC, Goldstein D, Summy D, Bart GD, Asphaug E, Korycansky D, Landis D, Sollitt L (2010) Detection of water in the LCROSS ejecta plume. Science 330:463-468 doi:10.1126/science.1186986

10. Fixsen DJ (2009) the temperature of the cosmic microwave background. Astrophys J 707:916 doi:10.1088/0004-637X/707/2/916

11. Hansen J, Johnson D, Lacis A, Lebedeff S, Lee P, Rind D, Russel G (1981) Climate impact of increasing atmospheric carbon dioxide. Science 213(4511):957-966

12. Haywood AM, Hill DJ, Dolan AM, Otto-Bliesner BL, Bragg F, Chan W-L, Chandler MA, Contoux C, Dowsett HJ, Jost A, Kamae Y, Lohmann G, Lunt DJ, Abe-Ouchi A, Pickering SJ, Ramstein G, Rosenbloom NA, Salzmann U, Sohl L, Stepanek C, Ueda H, Yan Q, Zhang Z (2013) Large-scale features of Pliocene climate: results from the Pliocene Model Intercomparison Project. Clim Past 9:191-209 doi:10.5194/cp-9-191-2013

13. Hemingway BS, Robie RA, Wilson WH (1973) Specific heats of lunar soils, basalt, and breccias from the Apollo 14, 15, and 16 landing sites, between 90 and 350 K. In: Proceedings of the fourth lunar science conference, Supplement 4 Geochemica et Cosmochimica Acta. New York: Peragmon Press. pp 2481-2487

14. Kaltenegger L, Sasselov D (2011) Exploring the habitable zone for Kepler planetary candidates. Astrophys J 736:L25 doi:10.1088/2041-8205/736/2/L25

15.   Keihm SJ (1984) Interpretation of the lunar microwave brightness temperature spectrum: easibility of orbital heat flow mapping. Icarus   60:568-589   doi:10.1016/0019-1035(84)90165-9 Publisher Full Text

16.   Kiehl JT, Trenberth KE (1997) Earth's annual global mean energy budget. Bull Am Meteorol Soc 78:197-208

17.   Kopp G, Lean JL (2011) A new, lower value of total solar irradiance: evidence and climate significance. Geophys Res Lett 38: L01706, doi:10.1029/2010GL045777 Lacis AA, Schmidt GA, Rind D, Ruedy RA (2010) Atmospheric CO_2: principal control knob governing Earth's temperature. Science 330: doi:10.1126/science.1190653 Lacis AA, Hansen JE, Russell GL, Oinas V, Jonas J (2013) The role of long-lived greenhouse gases as principal LW control knob that governs the global surface temperature for past and future climate change. Tellus B 65:19734 doi:10.3402/tellusb.v65i0.19734

18.   Leconte J, Forget F, Charnay B, Wordsworth R, Selsis F, Millour E, Spiga A (2013) 3D climate modeling of close-in land planets: circulation patterns, climate moist bistability, and habitability. Astron Astrophys 554:A69 doi:10.1051/0004-6361/201321042

19.   Leyrat C, Coradini A, Erard S, Capaccioni F, Capria MT, Drossart P, De Sanctis MC, Tosi F, the VIRTIS Team (2011) Thermal properties of the asteroid (2867) Steins as observed by VIRTIS/Rosetta. Astron Astrophys 531:A168 doi:10.1051/0004-6361/201116529

20.   Lissauer JJ, Pater I (2013) Fundamental planetary science: physics, chemistry and habitability. New York NY: Cambridge University Press.

21.   Loeb NG, Wielicki BA, Doelling DR, Kato S, Wong T, Smith GL, Keyes DF, Manalo-Smith N (2009) Toward optimal closure of the Earth's top-of-atmosphere radiation budget. J Clim 22:748-766 doi: 10.1175/2008JCLI2637 Publisher Full Text Lunt DJ, Haywood AM, Schmidt GA, Salzmann U, Valdes PJ, Dowsett HJ (2010) Earth system sensitivity inferred from Pliocene modelling and data. Nat Geosci 3:60-64 doi:10.1038/ngeo706

22.   Mallama A, Wang D, Howard RA (2002) Photometry of Mercury from SOHO/LASCO and Earth: the phase functions from 2 to 170°. Icarus 155(2):253-264 doi:10.1006/icar.2001.6723

23. Manabe S, Möller F (1961) On the radiative equilibrium and heat balance of the atmosphere. Mon Weather Rev 89:503-532 doi:10.1175/1520-0493(1961)089<0503:OTREAH>2.0.CO;2

24. Manabe S, Strickler RF (1964) Thermal equilibrium of the atmosphere with a convective adjustment. J Atmos Sci 21:361-385   doi:10.1175/1520-0469(1964)021<0361:TEOTAW>2.0.CO;2

25. Marshall J, Plumb RA (2008) Atmosphere, ocean and climate dynamics: an introductory text. Massachusetts: Elsevier Academic Press.

26. Matthews G (2008) Celestial body irradiance determination from an underfilled satellite radiometer: application to albedo and thermal emission measurements of the moon using CERES. Appl Opt 47(27):4981-4993 doi:10.1364/AO.47.004981

27. Möller F (1964) Optics of the lower atmosphere. Appl Optics 3:157-166 doi:10.1364/AO.3.000157

28. Ochsner TE, Sauer TJ, Horton R (2007) Soil heat storage measurements in energy balance studies. Agron J 99:311-319 doi:10.2134/agronj2005.0103S

29. Paige DA, Foote MC, Greenhagen BT, Schofield JT, Calcutt S, Vasavada AR, Preston DJ, Taylor FW, Allen CC, Snook KJ, Jakosky BM, Murray C, Soderblom LA, Jau B, Loring S, Bulharowski J, Bowles NE, Thomas IR, Sullivan MT, Avis C, De Jong EM, Hartford W, McClees DJ (2010) The lunar reconnaissance orbiter diviner lunar radiometer experiment. Space Sci Rev 150:125-160 doi:10.1007/s11214-009-9529-2

30. Paige DA, Siegler MA, Zhang JA, Hayne PO, Foote EJ, Bennett KA, Vasavada AR, Greenhagen BT, Schofield JT, McCleese DJ, Foote MC, De Jong EM, Bills BG, Hartford W, Murray BC, Allen CC, Snook KJ, Soderblom LA, Calcutt S, Taylor FW, Bowles NE, Bandfield JL, Elphic RC, Ghent RR, Glotch TD, Wyatt MB, Lucey PG (2010) Diviner lunar radiometer observations of cold traps in the Moon's south polar region. Science 330:479-482

31. Peixoto JP, Oort AH (1992) Physics of climate. New York: Springer.

32. Pierrehumbert R (2010) Principles of planetary climate. New York: Cambridge University Press.

33. Pierrehumbert R (2011) Infrared radiation and planetary temperature. Phys Today 64:33-38

34. Presley MA, Christensen PR (1997) Thermal conductivity measurements of particulate materials 2. Results J Geophys Res 102(E3):6551-6566 doi:10.1029/96JE03303

35. Ramanathan V, Inamdar A (2006) the radiative forcing due to clouds and water vapor. In: Kiehl JT, Ramanthan V (eds) Frontiers of climate modeling, Cambridge: Cambridge University Press. pp 119-151

36. Rubincam DP (2004) Black body temperature, orbital elements, the Milankovitch precession index, and the Seversmith psychroterms. Theor Appl Climatol 79:111-131 doi:10.1007/s00704-004-0056-5

37. Russell GL, Lacis AA, Rind DH, Colose C, Opstbaum RF (2013) Fast atmosphere–ocean model runs with large changes in CO_2. Geophys Res Lett 40(21):5787-5792 doi:10.1002/2013GL056755

38. Schmidt GA, Ruedy R, Miller RL, Lacis AA (2010) the attribution of the present-day total greenhouse effect. J Geophys Res 115:D20106 doi:10.1029/2010JD014287

39. Schulze-Makuch D, Méndez A, Fairén AG, von Paris P, Turse C, Boyer G, Davila AF, António MR, Catling D, Irwin LN (2011) A two-tiered approach to assessing the habitability of exoplanets. Astrobiology 11(10):1041-1052 doi:10.1089/ast.2010.0592

40. Shestopalov DI, Golubeva LF (2011) Bond albedo of asteroids from polarimetric data. 42nd lunar and planetary science conference, Abstract # 1028. http://www.lpi.usra.edu/meetings/lpsc2011/pdf/1028.pdf *webcite*Smith A (2008) Proof of the atmospheric greenhouse effect. Atmospheric and oceanic physics. arXiv:0802.4324v1 [physics.ao-ph] (http://arxiv.org/PS_cache/arxiv/pdf/0802/0802.4324v1.pdf*webcite*) Spudis PD, Bussey BJ, Baloga SM, Butler BJ, Carl D, Carter LM, Chakraborty M, Elphic RC, Gillis–Davis JJ, Goswami JN, Heggy E, Hillyard M, Jensen R, Kirk RL, LaVallee D, McKerracher P, Neish CD, Nozette S, Nylund S, Palsetia M, Patterson W, Robinson MS, Raney RK, Schulze RC, Sequeira H, Skura J, Thompson TW, Thomson BJ, Ustinov EA, Winters HL (2010) Initial results for the north pole of the Moon from Mini-SAR, Chandrayaan-1 mission. Geophys Res Lett 37:L06204 doi: 10.1029/2009GL042259

41. Stephens GL, Li J, Wild M, Clayson CA, Loeb N, Kato S, L'Ecuyer T, Stackhouse PW Jr, Lebsock M, Andrews T (2012) An update on

Earth's energy balance in light of the latest global observations. Nat Geosci 5:691-696 doi:10.1038/ngeo1580

42. Trenberth KE, Fasullo JT, Kiehl J (2009) Earth's global energy budget. B Am Meteorol Soc 90:311-323 doi:10.1175/2008BAMS2634.1

43. Vasavada AR, Paige DA, Wood SE (1999) Near-surface temperatures on Mercury and the Moon and the stability of polar ice deposits. Icarus 141:179-193 doi:10.1006/icar.1999.6175

44. Vasavada AR, Bandfield JL, Greenhagen BT, Hayne PO, Siegler MA, Williams JP, Paige DA (2012) lunar equatorial surface temperatures and regolith properties from the diviner lunar radiometer experiment. J Geophys Res 117:E00H18 doi:10.1029/2011JE003987

45. Vondrak R, Keller J, Chin G, Garvin J (2010) Lunar Reconnaissance Orbiter (LRO): observations for lunar exploration and science. Space Sci Rev 150:7-22 doi:10.1007/s11214-010-9631-5

46. Wallace JM, Hobbs PV (2006) Atmospheric science: an introductory survey. California: Academic.

47. Wild M, Folini D, Schär C, Loeb N, Dutton EG, König-Langlo G (2013) The global energy balance from a surface perspective. Clim Dyn 40:3107-3134 doi:10.1007/s00382-012-1569-8

48. Williams DR (2014) NASA Planetary Factsheet. http://nssdc. gsfc.nasa.gov/planetary/factsheet/ *webcite* URL last accessed on Oct 20, 2014 Zeng X (2010) What is the atmosphere effect on the earth's surface temperature? Eos Trans 91(15):134-135 doi:10.1029/2010EO150002

49. Zent AP, Hecht MH, Cobos DR, Wood SE, Hudson TL, Milkovich SM, DeFlores LP, Mellon MT (2010) Initial results from the thermal and electrical conductivity probe (TECP) on Phoenix. J Geophys Res 115:E00E14 doi:10.1029/2009JE003420

Chapter 3

The Non-hydrostatic Icosahedral Atmospheric Model: Description and Development

Masaki Satoh[1, 2], Hirofumi Tomita[2, 3], Hisashi Yashiro[3], Hiroaki Miura[2, 3, 4], Chihiro Kodama[2], Tatsuya Seiki[2], Akira T Noda[2], Yohei Yamada[1, 2], Daisuke Goto[5], Masahiro Sawada[1], Takemasa Miyoshi[3], Yosuke Niwa[6], Masayuki Hara[2], Tomoki Ohno[1], Shin-ichi Iga[3], Takashi Arakawa[2, 7], Takahiro Inoue[2, 7], and Hiroyasu Kubokawa[1]

[1]Atmosphere and Ocean Research Institute, The University of Tokyo, 5-1-5 Kashiwanoha, Kashiwa 277-85648, Chiba, Japan

[2]Japan Agency for Marine-Earth Science and Technology, 3173-15, Showa-machi, Kanazawa-ku, Yokohama 236-0001, Kanagawa, Japan

[3]RIKEN Advanced Institute for Computational Science, 7-1-26, Minatojima-minami-machi, Chuo-ku, Kobe 650-0047, Hyogo, Japan

[4]Department of Earth and Planetary Science, The University of Tokyo, 7-3-1 Hongo, Bunkyo-ku 113-0033, Tokyo, Japan

[5]National Institute for Environmental Studies, 16-2 Onogawa, Tsukuba 305-8568, Ibaraki, Japan

[6]Meteorological Research Institute, 1-1 Nagamine, Tsukuba 305-0052, Ibaraki, Japan

[7]Research Organization for Information Science and Technology, 2-32-3, Kitashinagawa, Shinagawa-ku 140-0001, Tokyo, Japan

ABSTRACT

This article reviews the development of a global non-hydrostatic model, focusing on the pioneering research of the Non-hydrostatic Icosahedral Atmospheric Model (NICAM). Very high resolution global atmospheric circulation simulations with horizontal mesh spacing of approximately O (km) were conducted using recently developed supercomputers. These types of simulations were conducted with a specifically designed atmospheric global model based on a quasi-uniform grid mesh structure and a non-hydrostatic equation system. This review describes the development of each dynamical and physical component of NICAM, the assimilation strategy and its related models, and provides a scientific overview of NICAM studies conducted to date.

REVIEW

Introduction

Diabatic heating due to the release of latent heat in deep convection is the primary heat source in the atmosphere, and it is interacted with the atmospheric general circulation, especially the tropical large-scale overturning circulations such as the Hadley and Walker circulations. Individual deep convective cells are associated with meso-scale circulations that have a horizontal scale of O (10 km), and an upward convective core, along with a horizontal scale of O (km). Until recently, since the horizontal resolution of the global climate models that have been used for future climate change projections has been O (100 km), such models require the use of cumulus parameterizations in order to incorporate the effects of deep convection instead of by explicitly resolving deep convective circulations. However, it is known that cumulus parameterizations significantly affect the results of climate model simulations and that they are the most ambiguous factor used in climate models (Randall et al. [2003]).

To overcome the above-mentioned cumulus parameterization issue, global non-hydrostatic models that utilize a horizontal mesh interval of O (km) for global atmospheric circulation simulations have been developed. Such models explicitly calculate deep convective circulations over the global domain without using cumulus parameterizations. At the grid-resolvable scale, water vapor is saturated into the liquid or ice phase of water in the upward flow field in order to form clouds and is eventually converted to rain and snow through cloud microphysics processes. In global non-hydrostatic models, clouds are spontaneously organized and the multi-scale structures of convective systems are reproduced over the global domain.

The Non-hydrostatic Icosahedral Atmospheric Model (NICAM) (Tomita and Satoh [2004]; Satoh et al. [2008]; Satoh [2013]) was first designed to be run with a horizontal mesh size approximately 3.5 km over the global domain by using the Earth Simulator (http://www.jamstec.go.jp/es/en/*webcite*) which was launched by the Japan Agency for Marine-Earth Science and Technology (JAMSTEC) in 2002. NICAM uses an icosahedral grid, as shown in Figure 1. Higher resolution grids are recursively subdivided from a coarser resolution grid. Hereinafter, we will refer to the grid division level as the g-level. The number of points, arcs, and triangles of the icosahedral grids with g-level l are given as follows:

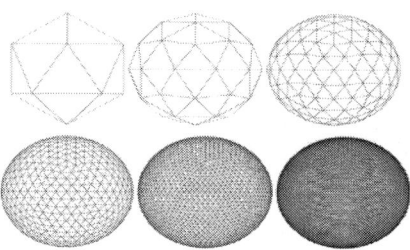

Figure 1: Icosahedral grids. Icosahedral grids for grid division levels of 0, 1, and 2 (top, from left to right), and 3, 4, and 5 (bottom, from left to right).

$$N_P = 10n^2 + 2 = 10 \times 2^{2l} + 2,$$

$$N_A = 30n^2 = 30 \times 2^{2l},$$

$$N_T = 20n^2 = 20 \times 2^{2l},$$

where $n = 2^l$. We define the average area of the triangles $\overline{A}$ and the average grid interval $\overline{\Delta}$ as follows:

$$\overline{A} = \frac{4\pi R^2}{N_T} = \frac{\pi R^2}{5 \times 2^{2l}},$$

$$\overline{\Delta} = \sqrt{2\overline{A}} = \sqrt{\frac{2\pi}{5}} \frac{R}{2^l},$$

where an Earth radius of $R = 6,371.22$ km is used. The values for each g-level are listed in Table 1.

Table 1: Number of points, arcs, triangles, and average grid interval of the icosahedral grids

	Points	Arcs	Triangles	Average area	Average interval
g-level	NP	NA	NT	$\overline{A}$ (km2)	$\overline{\Delta}$ (km)
0	12	32	20	5,100,996.991	7,142.126
1	42	122	80	1,275,249.248	3,571.063
2	162	482	320	318,812.312	1,785.532
3	642	1,922	1,280	79,703.078	892.766
4	2,562	7,682	5,120	19,925.769	446.383

5	10,242	30,722	20,480	4,981.442	223.191
6	40,962	122,882	81,920	1,245.361	111.596
7	163, 842	491,522	327,680	311.340	55.798
8	655, 362	1,966,082	1,310,720	77.835	27.899
9	2, 621,442	7,864, 322	5,242,880	19.459	13.949
10	10, 485,762	31, 457,282	20,971,520	4.865	6.975
11	41, 943,042	125, 829,122	83,886,080	1.216	3.487
12	167, 772,162	503, 316,482	335,544,320	0.304	1.744
13	671, 088,642	2,013, 265,922	1,342,177,280	0.076	0.872
14	2,684 ,354,562	8,053, 063,682	5,368,709,120	0.019	0.436

The g-level is the grid division level. Here, R = 6371.22 km.

Satoh et al.

Satoh et al. Progress in Earth and Planetary Science 2014 1:18, doi:10.1186/s40645-014-0018-1

NICAM has been shown to reproduce a realistic multi-scale cloud structure from a meso-scale to a planetary-scale cloud organization that is associated with the Madden-Julian Oscillation (MJO) (Madden and Julian [1971], [1972]) at a g-level between 9 ($\overline{\Delta}$ = 14 km) and 11 ($\overline{\Delta}$ = 3.5 km) (Tomita et al. [2005]; Miura et al. [2007b]). By using the K computer, which is installed at the RIKEN Advanced Institute for Computational Science (AICS) in Kobe, Japan (http://www.nsc. riken.jp/index-eng.html *webcite*), the resolution of NICAM has been recently increased to the subkilometer level, and it was shown that the deep convective core is more realistically resolved by using a g-level 13 ($\overline{\Delta}$ = 870 m) mesh simulation (Miyamoto et al. [2013]; Figure 2). In the current study, various experiments including decadal continuous experiments and case sweep experiments (Miyakawa et al. [2014]) were also conducted at g-levels between 9 and 11. Future projection studies such as those investigating changes in clouds and tropical cyclone activities are also investigated. The results are interpreted based on the more physically based cloud microphysics processes without the ambiguities of cumulus parameterizations.

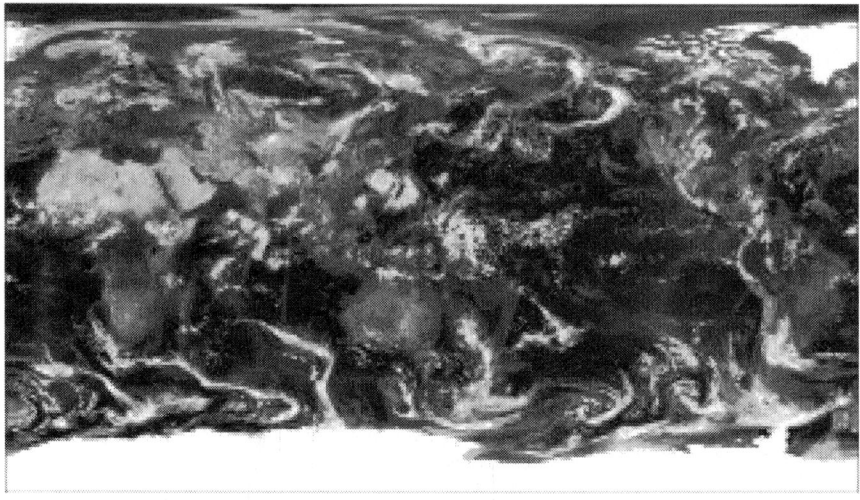

Figure 2: Cloud distribution simulated by the NICAM 870 m grid spacing experiment for 6:00 UTC 25 August 2012 (Miyamoto et al. [2013]).

This article describes the current development status, design, and concepts behind the individual components of NICAM. First, the background and an overview of studies related to the global non-hydrostatic model are reviewed. Then, the history of NICAM development and the scientific outcomes are summarized in the 'Scientific overview of NICAM' section, while the NICAM computational design is described in the 'Design, structure, development, and timeline' section. The two sections that follow describe the dynamical and physical components of NICAM, and their respective subsections describe each component of the physical processes. Next, the assimilation strategy is described. Finally, various NICAM usages that have been developed by modifying the original NICAM geometry are presented.

Review of Global Non-hydrostatic Models

Thanks to the significant advances in high-performance computers over the last decade, global atmospheric simulations with a horizontal resolution of O (10 km) can be achieved (Ohfuchi et al. [2004]; Mizuta et al. [2005]; Kinter et al. [2013]; Wedi [2014]). At enhanced horizontal resolution of less than 10 km, traditional atmospheric general circulation models (AGCMs) encounter fundamental difficulties in their dynamic framework formulation as well as in their computational efficiency. As their resolutions increase, AGCMs can capture flow features with comparable scales of motion in the horizontal and vertical directions (such as deep convection and fine-scale gravity waves) that can invalidate the hydrostatic approximation. Although deep convective systems in the tropics play key roles in global atmospheric circulations, they have not been directly resolved by AGCMs, and their effects have only been considered in a parameterized form (Arakawa [2004]). The effects of fine-scale gravity waves that are parameterized as gravity wave drag in AGCMs can be captured as the resolution increases, but the propagation of such gravity waves will be incorrectly calculated unless the non-hydrostatic effect is taken into consideration (Iwasaki et al. [1989]).

Regarding the numerical algorithm, many of the existing AGCMs employ the spectral transform method to represent spherical fields, and it has been pointed out that spectral transforms become increasingly inefficient for high-performance computing as the horizontal resolution increases (e.g., Stuhne and Peltier [1996]; Taylor et al. [1997]; Randall et al. [2000]; Satoh et al. [2005]; Cheong [2006]; Tomita et al. [2008]; Wedi [2014]). Another problem that occurs during computation on a massively parallel computer is that the spectral transform method requires extensive data movement between computer nodes. Although the double Fourier transformation method has been proposed as an alternative (e.g., Cheong [2006]), this method still requires global communication between computer nodes.

To increase the horizontal resolution beyond O (10 km) in a global atmospheric model, the governing equations and numerical algorithms must be reconsidered. More specifically, the governing equations must be non-hydrostatic, and the grid point method replaces the spectral method. As for the familiar latitude-longitude grid (lat-lon grid), however, the grid spacing near the poles becomes drastically reduced as the horizontal resolution is increased, which means that reduced grids are generally required to avoid severe time interval restrictions for the Courant-Friedrichs-Lewy (CFL) condition. In principle, the semi-Lagrangian, semi-implicit (SLSI) approach (cf. Laprise [2008]; Staniforth and Wood [2008]) could be employed to overcome the requirement of the CFL condition. Several authors (Semazzi et al. [1995]; Cullen et al. [1997]; Qian et al.[1998]; Côté et al. [1998]; Yeh et al. [2002]; Davies et al. [2005]; Wedi and Smolarkiewicz[2009]; Wood et al. [2013])

have used the lat-lon grid to solve a set of non-hydrostatic equations using the SLSI approach in order to acquire a larger time interval for integration. However, it is unclear how effective the elliptic solvers, which were developed for SLSI schemes, would be for ultra-high-resolution calculations, even though it has been recently proven that a multi-grid approach is an ideal solver for massively parallel computers (Heikes et al. [2013]).

The pole problem can be overcome by using grid systems with quasi-homogeneous grids over the sphere. One such grid is the icosahedral grid, which is currently one of the major grid systems used for high-resolution global atmospheric modeling. Primitive (hydrostatic) equation global models using icosahedral grids have been developed at Colorado State University for climate modeling (CSU AGCM) (Ringler et al. [2000]), Deutscher Wetterdienst for numerical prediction modeling (GME) (Majewski et al. [2002]), the National Oceanic and Atmospheric Administration (NOAA) for the Flow-following finite-volume Icosahedral Model (FIM) (http://fim. noaa.gov/*webcite*), and the Laboratoire de Météorologie Dynamique (DYNAMICO) (http://www.lmd.polytechnique.fr/~dubos/DYNAMICO/ *webcite*). Global non-hydrostatic models using icosahedral grids are also being developed by several international groups including the geodesic grid model at Colorado State University (UZIM;http:// www.cmmap.org/research/models.html *webcite*), the Icosahedral Non-hydrostatic model (ICON) at the Deutscher Wetterdienst and the Max Planck Institute for Meteorology (http://www.mpimet.mpg. de/en/science/models/icon.html *webcite*; Zängl et al, [2014]), the Model for Prediction Across Scales (MPAS) at the National Center for Atmospheric Research (Skamarock et al. [2012]), and the Non-hydrostatic Icosahedral Model (NIM) at NOAA (http://www.esrl.noaa. gov/gsd/ab/ac/GPU_Parallelization_NIM.html *webcite*). Entries for all these models can also be found at https://www.earthsystemcog. org/projects/dcmip-2012/*webcite*. As for other types of grid models, cubic grids are also candidates for high-resolution atmospheric models (McGregor [1996]; Lin [2004]; Putman and Suarez [2011]).

In Japan, the icosahedral atmospheric model has been created by using a non-hydrostatic system, i.e., NICAM (Tomita and Satoh [2004]; Satoh et al. [2008]; Satoh [2013]; http://nicam.jp/*webcite*). Development of NICAM began around 2000. Since high-resolution modeling has now entered the mainstream worldwide, NICAM has joined a number of international high-resolution numerical modeling projects, such as the Athena Project (Kinter et al. [2013]) with the Integrated Forecast System (IFS) (Jung et al. [2012]) and the Icosahedral-grid Models for Exascale Earth system simulations (ICOMEX) (Zängl et al. [2011]) with ICON, MPAS, and DYNAMICO.

Scientific Overview of NICAM

While more than a decade has passed since the development of NICAM began, the first milestone was reached in 2004 when Tomita and Satoh ([2004]) began finalizing the dynamical core of the model. Two unique characteristics of the NICAM dynamical core are the use of the spring dynamics to construct a modified icosahedral grid (Tomita et al. [2001], [2002]) and the use of a flux form non-hydrostatic system which guarantees conservation of mass and energy (Satoh [2002],[2003]). Tomita et al. ([2001], [2002]) constructed NICAM's icosahedral grids by improving the numerical accuracy of differential operators and using spring dynamics to improve the homogeneity of the grid system. An improved version of the spring dynamics method, which is more homogeneous and applicable even for higher resolution, is also now available (Iga and Tomita [2014]). NICAM uses a fully compressible (elastic) non-hydrostatic system to obtain statistically equilibrium states by performing long-timescale simulations. For this purpose, we devised a non-hydrostatic numerical scheme that guarantees the conservation of mass and energy (Satoh [2002], [2003]). Tomita and Satoh ([2004]) implemented this non-hydrostatic scheme in their global model using an icosahedral grid configuration, with which they developed a dynamical core for NICAM. Since the split-explicit time integration scheme (Klemp and Wilhemson [1978]) is used for the horizontal propagation of fast waves and for the implicit treatment of the vertical propagation of fast waves, multi-dimensional elliptic solvers are not required. To extend the original non-hydrostatic scheme to the global domain, the set of equations provided in Satoh ([2002], [2003]) is reformulated for spherical geometry and modified in order to make the equations suitable for an icosahedral grid configuration. The finite volume method is used for numerical discretization, so that the total mass and energy over the domain are conserved. The resulting model is suitable for long-term climate simulations.

NICAM is used for 'cloud-resolving simulations' by explicitly resolving convective circulation. This is accomplished by drastically increasing the horizontal resolution using high-power computing systems, such as Earth Simulator and the K computer. After finalizing the development of the dynamical core (Tomita and Satoh [2004]), the first 3.5 km mesh global NICAM simulation was performed on Earth Simulator by Tomita et al. ([2005]) using the aqua planet configuration (Neale and Hoskins [2001]). The results of this simulation clearly show multi-scale convective systems propagating near the equator (Satoh et al. [2005]), which is similar to the propagation of the observed cloud clusters and super cloud cluster structure (Hayashi and Sumi [1986]; Nakazawa[1988]; Takayabu et al. [1999]).

Tomita et al. ([2005]) conducted aqua planet experiments at three horizontal resolutions, g-levels 9, 10, and 11 ($\overline{\Delta}$ = 14, 7, 3.5 km), using the same physical schemes without cumulus parameterization. In their results, the multi-scale convective structure along the equatorial zone and the diurnal cycle of convective precipitation were reproduced similarly in all simulations, although the propagation speed along the equator and the phase lag of the diurnal cycle depends on the resolution. This result motivated us to use a combination of different horizontal resolution experiments with the same physical schemes when studying convective properties simulated by NICAM.

The results of the NICAM aqua planet experiment were analyzed intensively by Nasuno et al. ([2007], [2008]), Nasuno ([2008]), and Nasuno and Satoh ([2011a], [b]), which showed the roles of the equatorial convective systems embedded in the multi-scale structure. Mapes et al. ([2008]) also investigated the persistent structure in the tropics in terms of the predictability of disturbances. The diurnal cycle of tropical convective precipitation was clearly simulated with the dependency of the phase lag on the horizontal resolution (Tomita et al. [2005]). The semi-diurnal cycle was also captured. This was further investigated by Yasunaga et al. ([2013]) by using the NICAM aqua planet experiment. The detailed structures of convective systems near the tropopause were analyzed by Kubokawa et al. ([2010]).

The aqua planet configuration is used to test cloud changes that are related to increased surface temperatures in order to imitate global warming conditions. Miura et al. ([2005]) compared the cloud cover responses between NICAM and the Model for Interdisciplinary Research on Climate (MIROC) (Hasumi and Emori [2004]), which is an ordinary resolution climate model (denoted as CCSR/NIES/ FRCGC in Miura et al. [2005]), under the aqua planet condition. The comparison results show that the cloud cover simulated by NICAM increases in high-latitude regions, while that by MIROC decreases. Such a contrasting response to cloud cover is very interesting and should be analyzed in more detail, as will be described later in this section. The change in the meridional distribution of relative humidity is shown and compared to the other model results by Sherwood et al. ([2010]).

The aqua planet experiments of NICAM were compared to the other model results in Blackburn et al. ([2013]), and additional aqua planet experiments were conducted by Yoshizaki et al. ([2012a], [b]) in order to investigate the multiple-scale convective structure along the equator.

A NICAM simulation that includes a realistic configuration with a land and sea contrast was performed by Miura et al. ([2007a]). The results of this simulation show not only realistically simulated tropical cyclogenesis but also a clear dependency of convective organization on a planetary boundary layer (PBL) scheme, specifically, water vapor transport. Using the simulated dataset, the diurnal variation of the convective systems over the Tibetan Plateau was analyzed by Sato et al. ([2007], [2008]).

A successful simulation of a realistic MJO event was presented by Miura et al. ([2007b]), in which a 1 month integration with a 7 km grid spacing and a 1 week integration with a 3.5 km grid spacing were performed. A large-scale cloud organization, its eastward propagation, and the multiple-scale structure of convective systems were realistically reproduced (Nasuno et al. [2009]). The similarity of the simulation to the observed MJO was further analyzed by Liu et al. ([2009]). In addition, multiple tropical cyclones were realistically generated from the active cloud areas of the MJO. Particularly, Fudeyasu et al. ([2008], [2010a], [2010b]) analyzed the evolution of a tropical storm in detail, which was generated from the MJO 2 weeks after the initial condition.

Thanks to the high-resolution numerical simulations performed by Miura et al. ([2007b]), simulation results can be analyzed or used in various ways. The dataset consists of 7 km mesh and 14 km mesh grid data for 1 month from 15 December 2006 and 3.5 km mesh grid data for 1 week from 25 December 2006. The diurnal cycle of tropical convective systems was intensively analyzed by Sato et al. ([2009]). Convective momentum transports of the MJO were analyzed by Miyakawa et al. ([2012]). The dataset was compared with the satellite observations of the Tropical Rainfall Measuring Mission (TRMM) and CloudSat (Masunaga et al. [2008]), the geostationary satellite, the Multi-functional Transport Satellite (MTSAT-1R) (Inoue et al. [2008]), and CloudSat and Cloud-Aerosol Lidar and Infrared Pathfinder Satellite Observations (CALIPSO) (Inoue et al.[2010]; Satoh et al. [2010]; Ham et al. [2013]), and with the estimation of cloud radiative forcing (Sohn et al. [2010]). The dataset was also used to develop a cloud parameterization for coarser resolution AGCMs (Watanabe et al. [2009]). The convective structure near the tropopause was also analyzed by Kubokawa et al. ([2012]). The energy spectrum and similar frequency spectra of the high-resolution data were analyzed by Terasaki et al. ([2009]) and Tsuchiya et al. ([2011]).

Following Tomita et al. ([2005]) and Miura et al. ([2007b]), studies of intra-seasonal variability (ISV) including MJOs and boreal summer intra-seasonal oscillations (BSISO) (Fu and Wang [2004]) are unique areas of NICAM studies. Various aspects of the MJO and tropical disturbances simulated by Tomita et al. ([2005]) and Miura et al. ([2007b]) are being analyzed as mentioned above. NICAM studies are also collaborated with observational field campaigns of MJOs. Miura et al. ([2009]) examined an onset of MJO observed at the Mirai Indian Ocean cruise for the Study of the MJO-convection Onset (MISMO) field campaign (Yoneyama et al. [2008]), suggesting a role of the sea surface temperature (SST) gradient in the western Indian ocean. Also, for the Cooperative Indian Ocean Experiment on Intraseasonal Variability in the Year 2011/Dynamics of the MJO (CINDY2011/DYNAMO) campaign (Yoneyama et al. [2013]), Nasuno ([2013]) summarized a result from quasi-real-time weather forecasting experiments that showed simulated characteristics of MJOs during the CINDY2011/DYNAMO period in the boreal winter between 2011 and 2012. The quasi-real forecasting system using the stretch NICAM (Tomita [2008a]; see 'Models related to NICAM' section) was developed particularly for use in collaboration with field experiments, as introduced by Oouchi et al. ([2012]). More statistical results for MJO simulations are shown by Miyakawa et al. ([2014]), who reported a good performance of MJO predictability by the hind cast approach with NICAM. BSISO simulations were analyzed by Oouchi et al. ([2009a], [2014]), Taniguchi et al. ([2010]), Yanase et al. ([2010b]), and Satoh et al. ([2012a]). In particular, they focused on relations between ISV and tropical cyclones, as described below, since ISV modulates large-scale vorticity distributions that generally lead to preferable conditions for cyclogenesis.

Tropical cyclones can also be adequately researched by using NICAM. Tropical cyclogensis and its evolution from the MJO were analyzed for boreal winter cases by Fudeyasu et al. ([2008], [2010a], [2010b]). For boreal summer, tropical cyclogenesis was found to be frequently related to ISV activities (Kikuchi and Wang [2010]). The cyclone Nargis, which caused severe damage to Myanmar in May 2008, was successfully simulated and its relation to the northward propagation of BSISO in the Bay of Bengal was analyzed by Taniguchi et al. ([2010]) and Yanase et al. ([2010a],[2012a]). Oouchi et al. ([2009a]) simulated tropical cyclogenesis in the northwestern Pacific and showed its relation to the MJO. The effects of an equatorial Kelvin wave on an abrupt development of a tropical cyclone were clearly shown by Yanase et al. ([2010b]). Yanase et al. ([2012b]) also analyzed some statistical behaviors of tropical cyclones simulated by NICAM especially over the eastern Pacific. The impacts of a tropical cyclone on the Baiu rainfall were analyzed by Yamaura et al. ([2013]). The preferable conditions of tropical cyclogenesis were analyzed over the Atlantic domain (Satoh et al. [2013]).

Tropical cyclone climatology and possible changes in the future projection of tropical cyclone activities are actively being researched using NICAM. For example, Satoh et al. ([2012a]) analyzed 8 years of boreal summer experiments that were conducted in the Athena Project (Kinter et al. [2013]) in order to show the simulated climatology of tropical cyclones and produced results that show good relations between tropical cyclone activities and MJO activities. Future changes in the tropical cyclone structure were analyzed by Yamada et al. ([2010], [2012]), Yamada and Satoh ([2013]), and Emanuel et al. ([2010]).

NICAM simulations also contribute to the studies on Asian monsoon and tropical convective systems. For example, Oouchi et al. ([2009b]) provided a realistic distribution of precipitation during a boreal summer and showed its variation for diurnal and intra-seasonal timescales. Additionally, the resolution dependency of the diurnal cycle of convective systems was analyzed by Sato et al. ([2009]), and their results were found to be generally consistent with those of the aqua planet experiment (Tomita et al. [2005]); as resolution becomes coarser, the timing of precipitation peak becomes increasingly delayed. Sato et al. ([2009]) also analyzed the diurnal evolution of tropical convective cold pools, while Fujita et al. ([2011]) analyzed the behavior of the diurnal convection over maritime continental regions. Dirmeyer et al. ([2012]) and Noda et al. ([2012]) also analyzed the diurnal cycle of precipitation over the global domain. A mechanism for the diurnal variation of summer precipitation over southern China was investigated by Satoh and Kitao ([2013]) using the stretched version of NICAM (see 'Models related to NICAM' section).

The use of a cloud microphysics scheme in NICAM experiments without cumulus parameterization leads to realistic behavior of cloud distribution in the horizontal and vertical directions. Furthermore, as seen in the paragraph below, it might even produce a different behavior of cloud feedback from that obtained by conventional climate models that use cumulus parameterization. Iga et al. ([2007a]) analyzed the climatological distribution of cloud cover and compared it with the data of the International Satellite Cloud Climatology Project (ISCCP) (Rossow and Schiffe [1999]), while in a later study, Iga et al. ([2011]) analyzed upper tropospheric ice clouds and their sensitivity using various model parameters. Noda et al. ([2010]) investigated the importance of the subgrid turbulent process on cloud distributions. Kodama et al. ([2012]) showed the sensitivities of upper clouds to the choice of various cloud microphysics parameters. Satoh and Matsuda ([2009]) investigated how cloud microphysics affects upper clouds under an idealized radiative convective equilibrium condition.

As for the future simulations of cloud change by NICAM, Miura et al. ([2005]) showed the model's different behavior when compared to a coarse resolution AGCM (MIROC). In addition, Collins and Satoh ([2009]) showed that as a response that was different from the existing models, the upper cloud cover might increase, whereas the ice water path decreases under future warming conditions. Satoh et al. ([2012b]) analyzed their results and speculated that a future reduction in convective circulation could lead to a reduction in the ice water path under the warming condition. Tsushima et al. ([2014]) further investigated longwave cloud feedback and showed the possibility of a stronger positive feedback than other coarse resolution AGCMs. Yamada and Satoh ([2013]) analyzed the relation between cloud changes over the global domain and tropical cyclone changes. Kodama et al. ([2014a]) also analyzed the cloud changes associated with storm tracks simulated by the NICAM aqua planet experiments.

Recently, use of the K computer has allowed the horizontal resolution of NICAM to be increased. Miyamoto et al. ([2013]) performed (for the first time) a subkilometer mesh global domain simulation with grid spacing of 870 m (g-level 13) and 96 vertical layers (Figure 2). They showed that at a resolution of less than 1.7 km, convective cores could be resolved by multiple grid cells. The ability to use the K computer introduces a new area of study for NICAM. Numerous MJO case studies are currently being conducted in order to evaluate their statistical predictability (Miyakawa et al. [2014]). More than 10 years of continuous simulations have and are being conducted under present and future conditions in order to investigate future changes to tropical cyclones and other extreme events. The subsequent sections describe the improvements in the simulated climatology that have been made possible by improving the physical processes.

As described above, NICAM is usually run without using cumulus parameterization. In general, the horizontal scale of meso-scale phenomena ranges from a few to several hundred kilometers. It is not readily understood why a model with horizontal mesh spacing of O (10 km) reproduces such meso-scale phenomena without subgrid convective parameterization. It is speculated that for relatively large-scale meso-scale convective systems, the statistical effects of circulation resolved by grids behave similarly to nature. Figure 3 shows resolution dependency of upper cloud distributions in the range of $\overline{\Delta}$ = 28, 14, 7, 3.5, 1.7, 0.87 km for the NICAM simulations conducted by Miyamoto et al. ([2013]). Although in that study, it was analyzed that the convective core becomes resolved at $\overline{\Delta}$ = 1.7 km. Figure 3 shows that the large-scale structure of upper clouds is nearly unchanged for these resolutions. This result implies that the usefulness of a model without cumulus parameterization depends on research targets of meteorological phenomena. Large-scale organized convective structures such as MJOs are, in general, well reproduced without using cumulus parameterization, even at the coarser horizontal resolutions around O (10 km) as also indicated by Holloway et al. ([2013]), while the precise simulation of details of meso-scale convective structure needs much finer resolutions (Bryan et al. [2003]).

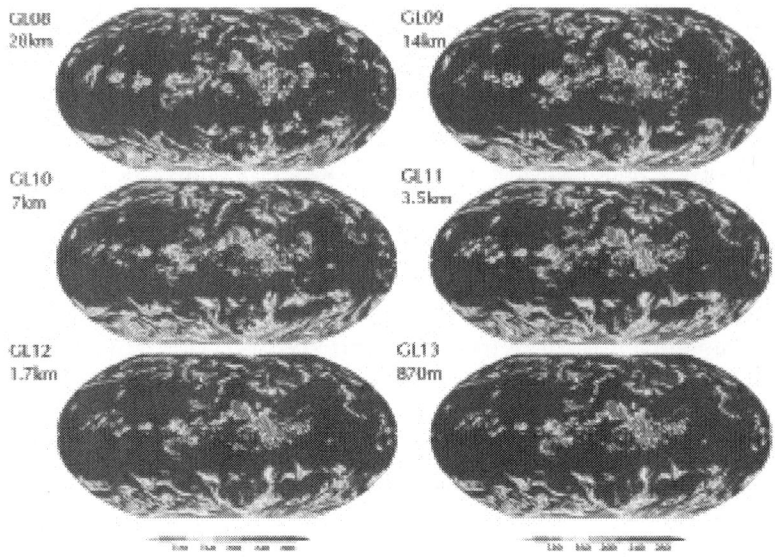

Figure 3: Resolution dependency of cloud structure simulated by NICAM between 28 km and 870 m grid spacing experiments. Cloud distributions simulated by NICAM between 28 km and 870 m grid spacing experiments for 6:00 UTC 25 August 2012, shown by the outgoing longwave radiation at the top of the atmosphere. The unit is W m^{-2}.

Design, Structure, Development, and Timeline

A timeline of the development of NICAM is summarized in Table 2. As was previously mentioned, the development of NICAM started in 2000 at JAMSTEC. Through the construction of icosahedral grid systems (Tomita et al. [2001], [2002]), a shallow water model (SWM) (Tomita et al. [2001], [2002]) was initially tested. At the same time, new non-hydrostatic numerical schemes were being developed (Satoh [2002], [2003]). By combining the two approaches, a three-dimensional (3D) dynamical core was eventually developed (Tomita and Satoh [2004]). The resulting 3.5 km mesh high-resolution dynamical core experiments were examined by Iga et al. ([2007b]). Then, full-physics experiments have been underway since 2004. Furthermore, the dynamical schemes, computational stability, and a tracer advection scheme (Miura [2007]; Niwa et al. [2011a]) have improved. For physical processes, we have actively introduced more sophisticated schemes, especially for cloud microphysics (Grabowski [1998]; Tomita [2008b]; Seiki and Nakajima [2014]) and PBL schemes (Nakanishi and Niino [2006], [2009]; Noda et al. [2010]). An accurate semi-Lagrangian scheme for sedimentation processes was developed for cloud microphysics schemes (Xiao et al. [2003]). Aerosol processes (Suzuki et al. [2008]) and atmospheric chemistry processes have also been introduced. Each component of the abovementioned physical schemes will be described in later sections of this review. Several NICAM-based systems are also currently under development, including atmosphere-ocean coupling and ensemble-based data assimilation systems, and downscaling methods for regional scale experiments have been developed for more advanced uses.

Table 2: Timeline of the development of NICAM

	2000	2001	2002	2003	2004	2005	2006	2007	2008	2009	2010	2011	2012	2013	2014
Grid system			→Modified grid using the spring dynamics				→Grid stretching method				→XTMS	→Diamond NICAM, plane NICAM			
Dynamics		→SWM		→3D dynamical core											
Basic scheme				→Non-hydrostatic scheme											
Advection scheme							→Miura scheme		→CWC consideration						
Physics															
Cumulus parameterizations model											→Tiedke	→Chikira			
Cloud microphysics model					→Grabowski		→NSW6			→NDW6					
Land surface model				→Bucket	→MATSIRO										
Ocean surface model				→Fixed SST		→Slab Ocean								→Coupled Ocean	
PBL model				→MY2.5						→MYNN					
Radiation model				→MSTRN-X											
Aerosol/chemistry						→SPRINTARS							→CHASER		
Frameworks										→NICAM-LETKF	→Coupler (Jcup) →NICAM-TM				
Supercomputers			→ES							→ES2			→K computer		
Milestone experiments					→Aqua-Planet Experiment	→MJO Experiment			→Athena Project					→Subkilometer Experiment	

See the 'Abbreviations' section for the expanded forms of the abbreviations used in this table.

Satoh et al.

Satoh et al. Progress in Earth and Planetary Science 2014 1:18, doi:10.1186/s40645-014-0018-1

Since high-resolution simulations require large computational resources, the efficiency of state-of-the art supercomputers needs to be maximized. In general, weather/climate models use a large number of physical variables during computation, so the size of 3D arrays significantly increases with increases in spatial resolution. Large-sized, various, frequent data output is also needed for simulations. Taken together, these factors make it clear that weather/climate models require faster data throughput in all layers of a computer system. NICAM has been tactically designed to achieve efficient throughput at the 1) memory layer, 2) network layer, and 3) file input/output (I/O) layer.

Since numerous simple arithmetic operations are used for large arrays in NICAM algorithms, memory throughput is the most important aspect governing total computational efficiency. For example, when using Earth Simulator, the ratio of memory bandwidth to the floating-point performance (Byte/Flops ratio; hereinafter B/F ratio) is 4, which makes it easy to execute NICAM with a high degree of efficiency. However, in the past decade, the B/F ratio has decreased during the recent evolution of supercomputers. As a result, when NICAM was implemented on the K computer at B/F = 0.5, numerous optimizations had to be applied to improve the efficiency of computer performance. Saving memory transfers is one effective optimization tactic. To accomplish this, we suppressed the memory copies, reduced the intermediate arrays, and avoided unnecessary zero-filling. These changes had to be performed manually and required labor-intensive efforts. However, as a result of these optimizations, we succeeded in reducing the model execution time by 30%.

To improve network throughput, the finite volume method with an explicit scheme was adopted to minimize the global communications, as described in the section 'Dynamics'. When 640 nodes on the K computer are used, the network communication of NICAM accounts for about 10 % of the total elapsed time at the case of g-level 9 ($\overline{\Delta}$ = 14 km). This percentage increases as the number of nodes increases. Kodama et al. ([2014b]) developed an algorithm for optimally assigning of nodes in a network topology such as a torus/mesh.

For the file I/O throughput, NICAM adopts a distributed file I/O. If the output data are aggregated into a single file at the time of output, it would be easy to handle the file. However, NICAM cannot maximize the I/O throughput using such a single output file. The data of each region on the icosahedral grid of NICAM are output to a file at the node, and the icosahedral grid is converted into a lat-lon grid during the post-process. Generally, we use a simple bilinear interpolation for the post-process, although we have an option of another interpolation to keep conservation of area integration. A recently developed post-process program can run in parallel with the main model using a coupler (see 'Coupler' section). However, an increase in file size due to the higher resolution is a current issue, and we are planning to change the file type to a compressible standard format, such as Network Common Data Form version 4 (NetCDF4) (http://www.unidata.ucar.edu/software/netcdf/ *webcite*). Recently, a benchmark run on the K computer has achieved 10% of the peak performance using five nodes (40 cores), and usage has been verified with 81,920 nodes (655,360 cores). Furthermore, a subkilometer mesh global domain simulation (Miyamoto et al. [2013]) was executed with a performance of 230 Tflops in double precision for 68 billion grid cells while using 20,480 nodes of the K computer.

Further information on the NICAM development team and the initial developments can be found athttp://nicam.jp *webcite*. NICAM team members consist of researchers belonging to several research institutions in Japan. The source code is primarily written in Fortran 95 and follows the Meteorological Research Institute/Japan Meteorological Agency (MRI/JMA) standard coding rule (Muroi et al. [2002]). The program is parallelized using Message Passing Interface (MPI) process parallelization via automatic compiler thread parallelization. OpenMP is not actively used as the current program. Each component is managed as one or more modules. In order to maintain readability of the source code, C preprocessor macros are not used. The source code has been managed at repositories using the Concurrent Versions System (CVS), and the NICAM development team is currently moving to multi-branch development using the Git distributed revision control and source code management system. Due to its flexible portability, NICAM can run on various computer platforms, such as Earth Simulator, the K computer, IBM AIX, Linux, and Mac OSX. We have also tested several combinations of MPI libraries and Fortran compilers.

NICAM is also used as a benchmark program or a test platform for new architectures. For this purpose, we have released the dynamical core of NICAM as open source software (NICAM-DC), which is distributed under the terms of the Berkeley Software Development (BSD) 2-Clause License and is available for download (http://scale.aics.riken.jp/nicamdc/ *webcite*). The NICAM development team also participates in a working group for the common base library environment of weather and climate models and in cooperative development efforts with other modeling groups in Japan. In the field of computational science, NICAM has made significant contributions as a benchmark program and is one of the main application programs used in Exascale computing research.

Dynamics

Grid Configuration and Advection Scheme

As reviewed in the 'Introduction' section, the use of grid models with finite-difference and finite-volume methods for weather and climate models was reconsidered as massively parallel computer architectures entered widespread use. An icosahedral grid is a globally quasi-uniform and is possibly the most suitable base for the development of an atmospheric model that can be run on a supercomputer. Additionally, the adoption of an icosahedral grid for the horizontal grid of NICAM was partially motivated by the results of Heikes and Randall ([1995]), in which a two-dimensional (2D) shallow water system model was developed using the Z-grid arrangement of the prognostic variables (Randall [1994]; Figure 4d), which followed Masuda and Ohnishi ([1986]). The Z-grid model, however, had a disadvantage in that the model needed to diagnose the velocity fields from the vorticity and divergence fields by solving two Poisson equations in each time step. Furthermore, it was not obvious that an efficient Poisson solver would become available for use on the massively parallel computers, at least when the NICAM development began.

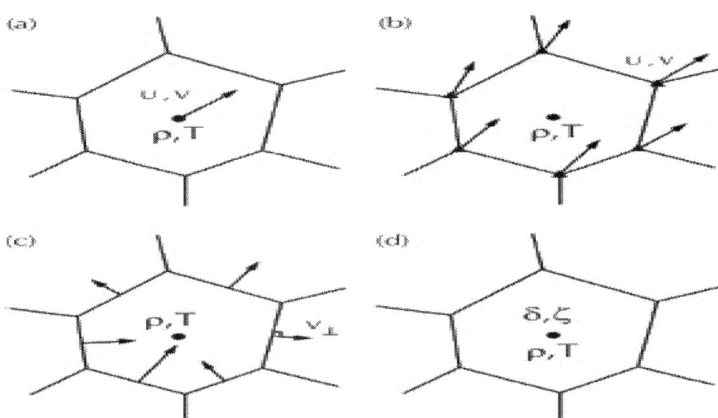

Figure 4: Arrangements of variables on the hexagonal grid.Arrangements of prognostic variables on the hexagonal grid for the (a)A-grid,

(b) B/ZM-grid, (c) C-grid, and (d) Z-grid staggering. The symbols ρ, T, u, v, $v_{\perp}$, δ, and ζ refer to density, temperature, zonal components, and meridional components of the flow velocity, respectively, that is defined (a) at the center or (b) at the vertex of the hexagon, normal component of the flow velocity on each edge, divergence of the horizontal flow, and vertical component of vorticity.

Tomita et al. ([2001]) did not follow any previous study (such as Heikes and Randall [1995]; Masuda and Ohnishi [1986]) and instead examined a new method by using the Arakawa A-grid (Arakawa and Lamb [1977]), which located prognostic variables (fluid depth and velocity) at the same nodes (Figure 4a) and thus did not require Poisson solvers. The accuracy and stability of the shallow water model were improved after applying a grid optimization that is now commonly known as 'spring dynamics' (Tomita et al. [2002]), and the physical performance of the model was evaluated using the standard tests provided in Williamson et al. ([1992]). The methods used to discretize the horizontal gradient, divergence, rotation, and Laplacian operators in NICAM are similar to those described in Tomita et al. ([2001]) and Tomita and Satoh ([2004]), except for some minor updates in the passive tracer transport (Miura [2007]; Niwa et al. [2011a]) and damping operators.

When the model was first developed by Tomita et al. ([2005]), NICAM was capable of producing a global simulation with horizontal grid spacing of 3.5 km. It was also the first model to produce a global simulation with horizontal grid spacing of less than 1 km (Miyamoto et al. [2013]). These characteristics indicate that the A-grid arrangement is a good choice due to its simplicity and the relative ease with which it attains high computational performance on Earth Simulator and the K computer. It should be noted that Tomita et al. ([2001]) provided the first example of the shallow water model on an icosahedral grid, which uses the finite-volume method and adopts velocity components as prognostic variables. However, it is well known that the A-grid model is unsuitable for simulations of inertia-gravity waves and geostrophic adjustments and that the model suffers from grid-scale noise (Randall [1994]). Therefore, other options should be considered for the horizontal variable arrangement, such as those described below.

The disadvantages of the B/ZM-grid (Figure 4b) and C-grid (Figure 4c) arrangements, which were pointed out by Ničkovic et al. ([2002]), on the hexagonal grid were partially solved by Ringler and Randall ([2002]) and Thuburn et al. ([2009]), respectively. Although problems with the computational modes remain due to the mismatch of the degrees of freedom between the fluid depth/pressure and horizontal velocity, the B/ZM-grid and C-grid not only allow a more realistic representation of inertia-gravity waves and geostrophic adjustments than the A-grid but also are free from the checkerboard pattern. Currently, we are investigating the B/ZM-grid as a candidate to replace the A-grid, and the difficulty in managing the computational mode has been largely eliminated by revising the operators of Ringler and Randall ([2002]). This activity is ongoing, and details of the improvements will be reported in forthcoming papers. The update of the dynamical core will also include the newly developed transport algorithms of Miura and Skamarock ([2013]) and Miura ([2013]).

Vertical Resolution Issues

NICAM adopts the Lorenz grid and z^* (terrain-following) coordinates (Gal-Chen and Somerville[1975]) as a vertical coordinate system (Figure 3 of Satoh et al. [2008]). Density, horizontal velocity, and internal energy are defined at the full levels, and vertical velocity is defined at the half levels. In NICAM, z^* is generally specified so as to linearly depend on the geometric height $z$between the model surface and top (the linear-z^* coordinate). Using the standard vertical levelsZ_k, geometric height $z_{X,k}$ is formulated as

$$z_{X,k} = Z_k + \left(1 - \frac{Z_k}{H_{top}}\right) Z_{X,sfc},$$

where k is an index for the vertical level, X represents the location of the horizontal grid, and $H_{top}$and $Z_{X,}sfc$ are the heights of the model top and bottom, respectively. In general, $H_{top} = 40$ km is used with 40 vertical levels. For standard cases, the lowermost full level is located at 81 m. The vertical interval of the half-level ΔZ_k increases with height, from $\Delta Z_k = 170$ m at the bottom, $\Delta Z_k < 1$ km below 10 km, and $\Delta Z_k \sim 3$ km near the model top.

In some situations, the hybrid-z* coordinate (Simmons and Burridge [1981]) is employed to reduce the amount of pressure gradient force numerical errors that arise from topography. It is formulated as

$$z_{X,k} = Z_k + \left(\frac{\sinh \frac{H_{top}-Z_k}{H_{efold}}}{\sinh \frac{H_{top}}{H_{efold}}} \right) Z_{X,sfc},$$

where He_{fold} is a height scale parameter (e.g., $He_{fold} = 10$ km). In the hybrid-z* coordinate, intervals of the vertical level approach those of the standard vertical level for $Z_k \gg He_{fold}$, whereas in the linear-z* coordinate, it is constant with height even near the model top. This hybrid-z* coordinate advantage is obtained at the expense of packing more vertical levels just above the higher mountain, which may cause the vertical CFL condition to become a serious issue for the hybrid-z* coordinate. Klemp ([2011]) also proposed a hybrid approach, in which the coordinate surfaces are smoothed with height in order to remove smaller scale terrain structures from the surfaces.

Recently, we constructed a new configuration for the vertical levels of NICAM in order to perform simulations with finer vertical resolutions in the troposphere, stratosphere, and mesosphere levels. The development of this 'middle atmosphere NICAM' is motivated by the results of recent climate studies in which the tropospheric circulations were found to be affected by the stratosphere. It should be noted that the inclusion of the middle atmosphere into a climate model is a worldwide trend in the climate model community, since it improves the reproducibility of the large-scale tropospheric circulation (e.g., Shindell et al. [1999]). Currently, the middle atmosphere NICAM employs the hybrid-z* coordinate, and its model top height is set to 80 km. The vertical intervals in the lower atmosphere are the same as the standard vertical coordinate. As the vertical interval increases with height and reaches a specific criteria (Δz_{max}), the vertical level is set with the uniform interval of Δz_{max} (except near the model top) in order to appropriately simulate the propagation of atmospheric gravity waves. We have tested $\Delta z_{max} = 2$, 1, and 0.5 km with a horizontal resolution of g-level 7 ($\overline{\Delta} = 120$ km), as shown in Figure 5. As the vertical resolution increases, the reproducibility of the zonal mean temperature and zonal wind tends to improve in the upper troposphere and middle atmosphere. Interestingly, westerly winds around the equatorial lower troposphere in June 2004 area only well simulated when the vertical resolution is set to 1 km or less. NICAM has an option that allows the deep atmosphere configuration (Tomita and Satoh[2004]) to be switched. For the middle atmosphere simulations, the choice of the deep atmosphere is considered appropriate.

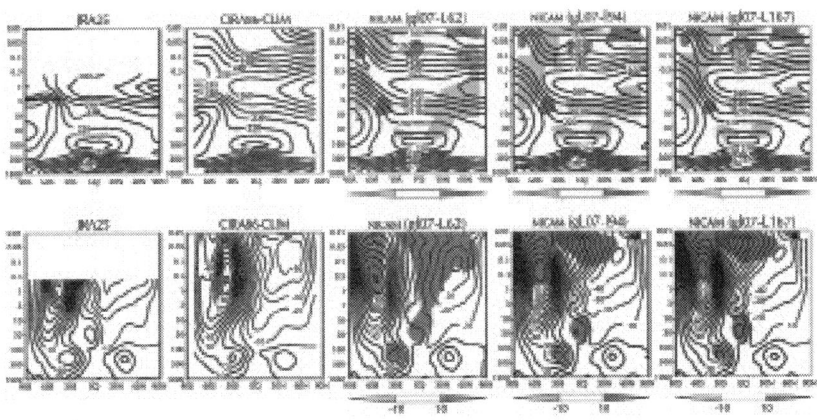

Figure 5: Dependency of the simulated NICAM climatology on the vertical resolution. Dependency on the vertical resolution of the (upper) zonal mean temperature (K), and (lower) zonal mean zonal wind (m s⁻¹). From left to right: JRA25 reanalysis (Onogi et al. [2007]), CIRA86 (http://badc.nerc.ac.uk/view/badc.nerc.ac.uk__ATOM__dataent_CIRA *webcite*), NICAM (Δz_{max} = 2 km), NICAM (Δz_{max} = 1 km), and NICAM (Δz_{max} = 0.5 km). For CIRA86, the climatology for June is shown. All the other panels are the averages for June 2004. Shading denotes deviations from the CIRA86 climatology.

Determining the proper vertical coordinates is an ongoing issue for high-resolution non-hydrostatic models. For global non-hydrostatic models in particular, the numerical instability of the terrain-following coordinates becomes more severe because the steep topography over the Andes or the Antarctica is within the simulation domain. For the present simulations, relatively coarser topographic data were used by applying a smoother for the NICAM simulations. For example, for the

3.5 km mesh simulations, 14 km mesh coarsened topographic data were generally used. In addition, a minimum threshold value of the Brunt-Väisälä frequency was specified for the calculations in the PBL schemes, that is and the Mellor-Yamada Nakanishi-Niino (MYNN) scheme, in order to suppress an unusual increase in vertical shears near the surface.

To overcome the problem of terrain-following coordinates, different candidates have been proposed. For example, the possibility of a vertical coordinate that is based on geopotential height, i.e., the z-coordinate, has been investigated. Although this approach is common for ocean models (Adcroft et al. [1997]), it must be understood that the implementation of the z-coordinate to non-hydrostatic atmospheric models is not straightforward (Gallus and Klemp [2000]). The immersed boundary method is well known for computational fluid dynamics (Mittal and Iaccarino [2003]). Various approaches to overcome the problems of the z-coordinate were taken into consideration, such as the cut-cell method or the thin-wall approximation (e.g., Bonaventura [2000]; Steppeler et al. [2002], [2006]; Yamazaki and Satomura [2008], [2010]), have been taken into consideration. We plan to introduce such an option for the z-coordinate of NICAM in the near future.

Physics

Cloud Microphysics Schemes

Cloud microphysics can be used to solve the growth equations of a single hydrometeor particle. The related processes are mainly categorized into four components: nucleation processes, phase change, collisional processes, and gravitational sedimentation. These processes have not been calculated explicitly and have been parameterized in conventional AGCMs since the characteristic timescale of these processes and spatial variability of cloud ensembles cannot be resolved using the time-step and grid resolution of GCMs. Global non-hydrostatic models operate by calculating these processes via cloud microphysics schemes in order to explicitly evaluate time evolution of cloud growth over the global domain. The implemented schemes are generally developed for regional non-hydrostatic models. This section introduces cloud microphysics models that are applicable to global non-hydrostatic models. For a more fundamental theoretical basis of cloud microphysics processes, refer to Pruppacher and Klett ([1997]).

Multi-Moment Bulk Method

Hydrometeor particles typically have a radius r from 10 μm to 1 mm and populations from 10^2 to 10^9 m^{-3}. To manage numerous particles efficiently (in terms of computational cost), cloud microphysics modeling generally assumes that particles with the same radius are homogeneously distributed within a grid cell of the model and develop in the same manner while under the same environmental conditions. Growth equations of hydrometeor particles with similar radii are solved together and the particle size distribution (PSD) $f(r)$, which is the number density sorted by the radius of the hydrometeors, is then predicted. Numerous observational studies have shown that liquid hydrometeor particles have three major modes for the PSD (Pruppacher and Klett [1997]): particles in the condensational growth mode (which have a mode radius around 10 μm), particles in the collisional growth mode (which have a mode radius larger than 100 μm), and the drizzling mode (which is a transitional mode between the two aforementioned modes).

There are two main methods that can be used to predict the time evolution of the PSD. One is the multi-moment bulk method, which predicts the integrand of each mode by approximating the modes via the gamma distribution (e.g., Milbrandt and Yau [2005b]; Seifert and Beheng [2006]):

$$f(r) = Kr^{\nu} \exp(-\lambda r^{\mu}),$$

where $\kappa, \lambda, \mu,$ and ν are the diagnosed parameters. The k-th moment of the PSD, $M^{(k)}$, is defined as

$$M^{(k)} \equiv \int_0^{\infty} f(r)r^k \, dr,$$

and the time evolution of the prognostic moments is evaluated by integrating the growth rates of a single hydrometeor particle over the entire range of the PSD (e.g., Seifert and Beheng [2006]). The other is the spectral bin method, which solves growth equations by discretizing the PSD with tens of bins covering these modes (Khain et al. [2000], [2008]). Because of their computational efficiency, the multi-moment bulk methods are widely used for 3D non-hydrostatic simulations.

The multi-moment bulk method is categorized by the number of its prognostic moments: the single-moment bulk method predicts the mass concentration of hydrometeors (Kessler [1969]; Lin et al. [1983]; Rutledge and Hobbs [1983]; Walko et al. [1995]; Grabowski [1998]; Hong et al.[2004]; Thompson et al. [2008]; Tomita [2008b]), and the double-moment bulk method predicts the number concentration in addition to the mass concentration (Meyers et al. [1997]; Feingold et al. [1998]; Morrison et al. [2005]; Seifert and Beheng [2006]; Phillips et al. [2007]; Morrison and Gettelman [2008]; Lim and Hong [2010]).

The four parameters used to represent the PSD (κ, λ, μ, and ν) are diagnosed by prognostic moments. Hence, as the number of the prognostic moments is increased, a variety of PSD shapes can be represented. For example, increases in the mode radius and skewness of the PSD are observed in collisional growth (Berry and Reinhardt [1974]). These distortions of the PSD are important for initiation of precipitation (Seifert and Beheng [2001]). Furthermore, once precipitation occurs, the gravitational sedimentation of the hydrometeor particles induces further distortion of the PSD due to the gravitational size sorting mechanism and the time evolution of the PSD affects the subsequent chain of particle growth (Wacker and Seifert [2001]; Milbrandt and Yau[2005a]; Milbrandt and McTaggart-Cowan [2010]). In addition, higher order moments are required for the closure of the governing equations in the collisional processes (e.g., Seifert and Beheng[2001]). Thus, an increase in the number of prognostic moments is linked to improvements in the time evolution of the PSD and complexity of the cloud microphysics modeling. Additionally, complexity and calculation costs increase as the number of prognostic moments increases.

For climate studies, collisional processes are particularly important to the radiative budget because such processes dominate the persistence of warm-phase clouds due to the rapid growth rate of cloud droplets during the precipitating process, along with the strong cloud albedo perturbations caused by the spatial and temporal variabilities of their growth rates (Albrecht [1989]; Lohmann and Feichter [2005]). The double-moment bulk method enables us to explicitly evaluate the dependence of the collisional growth rate on the mode radii (Berry and Reinhardt [1974]; Feingold et al. [1998]; Khairoutdinov and Kogan [2000]; Seifert and Beheng [2001]). In addition, cloud radiative forcing can be more accurately estimated because the cloud optical properties are calculated using the mass concentration and number concentration (Hansen and Travis [1974]). In contrast, the single-moment bulk method assumes that the radii of the hydrometeors are empirical values. Empirical values do not reproduce the temporal and spatial variability of atmospheric states, and their validity has not been confirmed for various types of climate change, such as global warming conditions. The triple-moment bulk method, which was recently developed and used (Milbrandt and Yau [2005a], [b]; Seifert [2008]; Shipway and Hill [2012]), predicts or diagnoses the radar reflectivity factor in addition to the aforementioned two moments and is applied to the regional simulations of severe storms (Milbrandt and Yau [2005a], [2006]). Of course, the higher order moment bulk methods also contain empirical formulations and parameters that can be used to calculate cloud growth and optical properties. However, previous process studies have shown that the differences of simulated results between double-moment and triple-moment bulk methods are smaller than those between single-moment and double-moment bulk methods (e.g., Milbrandt and Yau [2005a]; Shipway and Hill [2012]). In addition, it has also been demonstrated that a well-optimized double-moment method can capture the characteristics of cloud growth simulated by a spectral bin method (Seifert and Beheng [2001]; Seifert et al. 2006; Seifert [2008]; Milbrandt and Yau [2005a], [2006]). These facts indicate that the double-moment bulk method is practical and applicable in terms of cost performance of numerical simulations.

Cloud Microphysics Schemes in NICAM

NICAM initially adopted a single-moment bulk cloud microphysics scheme that was proposed by Grabowski ([1998]) in order to explore the multi-scale interaction of moist convective systems because the organizations of deep convective systems do not depend on the details of cloud microphysics schemes. Instead, meso-scale circulations are reproduced if the latent heat release in updrafts and the evaporation cooling caused by rain drops are considered by predicting the mass concentration of hydrometeor particles, as introduced in the single-moment bulk scheme. In the latest version, the schemes proposed by Kessler ([1969]), Lin et al. ([1983]), Hong et al. ([2004]), and Tomita ([2008b]) are implemented. In particular, NICAM single-moment scheme with six water categories proposed by Tomita ([2008b]) (hereinafter, referred to as NSW6) has been well evaluated by comparisons with satellite observations (Satoh et al. [2010]; Kodama et al.[2012]; Hashino et al. [2013]; Roh and Satoh [2014]) and, hence, is used by default. Because of the relatively low calculation cost, the use of single-moment schemes enables us to conduct very high-resolution global non-hydrostatic simulations at g-level 13 ($\overline{\triangle}$ = 870 m; Miyamoto et al.[2013]), or for longer integrations up to several decades.

Recently, Seiki and Nakajima ([2014]) implemented a double-moment bulk cloud microphysics scheme (NICAM double-moment scheme with six water categories; hereinafter, referred to as NDW6) into NICAM. Seiki et al. ([2014]) demonstrated that the NDW6 scheme well reproduces cloud radiative forcing by calculating cloud ice number concentration with using accurate databases of the cloud optical properties. Figure 6 shows the simulated cloud optical thickness (COT) and outgoing shortwave radiative flux at the top of the atmosphere that were obtained using the NDW6 scheme, along with the observed outgoing shortwave flux at the top of the atmosphere that was obtained from the satellite product of CERES SYN1deg_Ed3A (Wielicki et al. [1996]; Kato et al. [2011]). The simulation showed that convective clouds with a high COT were scattered over the tropics and formed regional cloud clusters (e.g., over the central Pacific). For example, Typhoon Fengshen remained over the Philippines and high COT clouds were distributed along its vortical structure. Furthermore, synoptic scale disturbances were organized over the mid-latitude regions and optically thick regions were located along the front. A shortwave flux corresponding to these high COT regions was largely reflected by clouds. These features agree well with the satellite observation. As shown by the NDW6 scheme, the impact of the cloud radiative forcing, which results in improvements to the climatological biases in NICAM, should be evaluated in future studies.

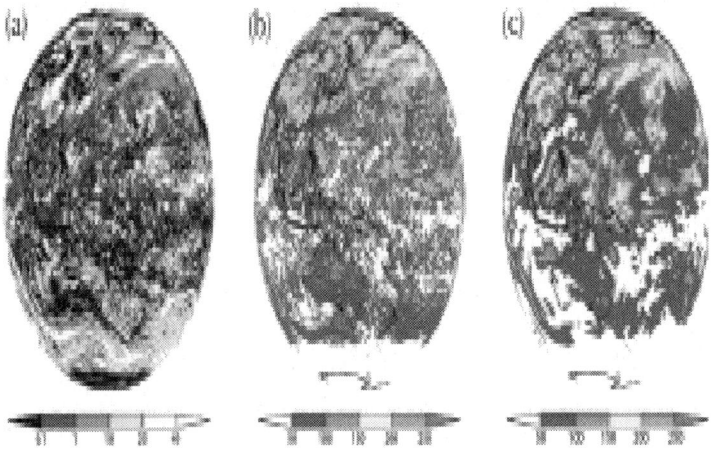

Figure 6: Simulated results for the 14 km mesh NICAM with the double-moment cloud microphysics scheme. Simulated results for the double-moment cloud microphysics scheme for the 14 km mesh NICAM on 20 June 2004: (a) simulated cloud optical thickness at a wavelength of 550 nm at 00:00 UTC, (b) daily averaged outgoing shortwave flux (W m^{-2}) at the top of the atmosphere, and **(c)** the daily averaged outgoing shortwave flux (W m^{-2}) at the top of atmosphere, as produced by the CERES SYN1deg_Ed3A product.

Subgrid-Scale Turbulence of the Planetary Boundary Layer

Boundary layer turbulence plays an important role in driving large-scale circulations of the atmosphere as it transports heat upward from the Earth's surface in the form of sensible and latent heats, which are converted from incoming solar energy at the surface (Trenberth et al. [2009]). To simulate atmospheric circulations, a parameterization scheme that computes the effects of such turbulent transport in the subgrid scale of a numerical model needs to be incorporated, unless the resolution is high enough to resolve turbulence.

NICAM adopts the MYNN scheme, which is a modified version of the original Mellor-Yamada (MY) model (Mellor and Yamada [1982]; Nakanishi and Niino [2006], [2009]), for such turbulent transport models (Noda et al. [2010]). There are three major advantages to employing the MY-type model: 1) the MY-type model can consistently and simultaneously compute subgrid-scale condensation and resultant turbulent fluxes by assuming the probability density functions regarding the inhomogeneity of temperature and water on a subgrid scale, 2) the model can consistently evaluate the impacts of the differences of turbulent moments, and 3) the model can run with low computational cost because it omits minor terms in the turbulent transport equation system. Based on the degree of approximation from the original equations, level 1 to level 4 can be included in the MY-type model. The level-2, level-2.5, and level-3 MYNN version models are installed in NICAM, and the level-2 model is used as a standard PBL scheme.

Figure 7 compares the impacts of these different level models on the spatial distributions of simulated clouds. As can be seen in the figure, all turbulent models properly simulate the spatial characteristics of, for example, high and mid-level clouds, even though the magnitude of the high clouds tends to be larger than those seen in satellite observations. For low clouds, the level-2 model properly represents the characteristics of the regional contrast between a clear and cloudy sky, such as those seen over the eastern and western regions of the Pacific Ocean. Conversely, the low clouds that systematically develop along and off the Peruvian coast are not simulated well. Compared with the results of the level-2, level-2.5 and level-3 models produce a much smaller fraction of low clouds. It is speculated that these two models calculate more vertical turbulent heat transports of heat from the boundary layer to the free atmosphere, which results in a less humid boundary-layer state.

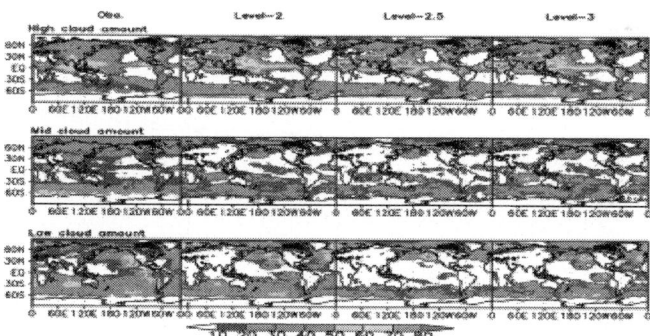

Figure 7: Comparison of the cloud amounts simulated by using different turbulent models to the satellite observation. The comparisons of the cloud

amounts from the ISCCP satellite data, level-2, level-2.5, and level-3 MYNN turbulent schemes were obtained by using the 14 km mesh NICAM. The NICAM data shown are averages during 29 July 1982 and 15 August 1982, and that of the satellite observation are the climatology for July and August.

Due to the remarkable progress of computational technology, the model resolution of NICAM is steadily being increased, with the highest resolution to date having been achieved when a horizontal mesh size of 870 m was used (Miyamoto et al. [2013]). For a global non-hydrostatic model, boundary-layer scale turbulence becomes explicitly and partially calculatable around this resolution, but will not be sufficiently resolved until the mesh size is less than 100 m (Khairoutdinov et al. [2009]; Moeng et al. [2009]). It is a challenging issue to parameterize turbulent transport processes in such an intermediate resolution (Wyngaard [2004]), and continuous efforts are needed to improve the subgrid-scale turbulence parameterization schemes along with shallow cumulus convection.

Radiation

Atmospheric radiative transfer models calculate heating rates in the atmosphere and radiative flux at the Earth's surface using the 3D distribution of clouds/gases/aerosols and land/ocean surface properties with the atmospheric state. NICAM adopts the broadband radiative transfer model 'MSTRN', which is based on the discrete ordinate method with a delta two-stream approximation and the correlated k-distribution method. The MSTRN model calculates gas absorption (H_2O, CO_2, O_3, N_2O, CH_4, and O_2) and the scattering/absorption of hydrometeors and aerosols (dusts, sea salts, carbonaceous, and sulfates) over an arbitrary time interval and then outputs short- and longwave radiation fluxes. A basic description of this model is written in Nakajima et al. ([2000]). In the latest package, 'MSTRN-X' (Sekiguchi and Nakaima [2008]), the gas absorption processes are improved and successfully used to reduce the error of the heating rate. The scheme was introduced into NICAM through the AGCM version implemented for MIROC from the original one-dimensional (1D) model. The calculation of solar insolation (Berger [1978]) was added to the AGCM version. The other cloud overlapping method (maximum-random overlapping) was incorporated since this version of the radiation scheme was included in MIROC3.2 (Hasumi and Emori [2004]).

The calculation cost of the radiation scheme is very high in the AGCM, because both computing speed and precision are required for the scheme. The latest MSTRN-X scheme improved the optimization method in order to reduce the integration points in the correlated k-distribution approximation without decreasing accuracy. In the NICAM default setting, the spectrum between 0.2 and 200 μm is divided into 29 spectral bands with 111 integration points. Cloud-radiation feedback is one of the most important issues in the study of climate change. The high-resolution experiments performed by NICAM have an advantage in that they reduce the uncertainties of the horizontal/vertical cloud distributions. In the recent studies described above, both the cloud scheme and radiation scheme are improved by considering the detailed size distribution of the hydrometeors and non-sphericity of the ice particles. This cooperative development of the cloud and radiation schemes will facilitate improvements in the future development of aerosol and chemistry modules (see 'Aerosol and chemistry modules' section).

NICAM also introduces the ISCCP simulator (Klein and Jakob [1999]; Webb et al. [2001]), with which the distribution of every cloud category can be calculated online. The results from the simulator have been used in numerous studies (Iga et al. [2007a]; Suzuki et al. [2008]; Collins and Satoh [2009]; Noda et al. [2010]; Sohn et al. [2010]; Satoh et al. [2012b]; Miyamoto et al.[2013]; Tsushima et al. [2014]). In some studies, new satellite simulator packages have been applied offline to the NICAM dataset. Kodama et al. ([2012]) used a series of satellite simulators that were part of the Cloud Feedback Model Intercomparison Project (CFMIP) Observation Simulator Package (COSP) (Bodas-Salcedo et al. [2011]), and Hashino et al. ([2013]) used the Joint Simulator for Satellite Sensors (J-simulator). Therefore, the online use of these packages should also be taken into consideration.

Aerosol and Chemistry Modules

Currently, NICAM is implemented with a global 3D aerosol transport-radiation model, the Spectral Radiation-Transport Model for Aerosol Species (SPRINTARS) (described fully in Takemura et al.[2000], [2002], [2005], [2009]; Goto et al. [2011a]). The SPRINTARS module predicts the mass mixing ratios of the main tropospheric aerosol species, i.e., carbonaceous (black carbon and organic matter), sulfate, soil dust, sea salt, and the precursor gases of sulfate (sulfur dioxide and dimethylsulfide (DMS)). The aerosol transport processes include emission, advection, diffusion, sulfur chemistry, wet deposition, dry deposition, and gravitational settling. Emissions of soil dust, sea salt, and DMS are calculated using the internal state of the model, whereas the emissions for the other primary aerosols and the distributions of oxidants (hydroxyl radical, ozone, and hydrogen peroxide) for the sulfur chemistry are prescribed by externally assumed data. In the module itself, meteorological fields such as clouds and precipitation, which are critical to determining aerosol lifetimes, are provided from the outer module after the cloud microphysics calculations are computed at each time step. In contrast, the calculated mass mixing ratios of the aerosols in each grid cell are calculated by the aerosol module SPRINTARS and then returned back to the outer module. In addition, aerosol variables are forwarded to the radiative transfer module, MSTRN-X (Sekiguchi and Nakaima [2008]), in order to represent the direct aerosol effects, and also to cloud schemes such as NSW6, to represent the indirect aerosol effects.

SPRINTARS, which was originally coupled to MIROC (Watanabe et al. [2010]) with a low spatial resolution of 100 to 300 km, has compared well with various measurements (Goto et al. [2011a],[b], [2012]) and other global aerosol models participating in the AeroCom international project (e.g., Kinne et al. [2006]). The results of the SPRINTARS module coupled to NICAM were also shown to generally agree with the satellite observations of aerosols and clouds for versions based on low spatial resolution of g-level 5 ($\overline{\Delta}$ = 220 km) (Dai et al. [2014a], [b]; Goto [2014]) and high spatial resolution of g-level 10 ($\overline{\Delta}$ = 7 km) (Suzuki et al. [2008]). Using Earth Simulator, Suzuki et al. ([2008]) conducted a global simulation with the finest resolution to date and showed the global aerosol and cloud distributions for July 2006 (Figure 8). Most of the aerosol plumes were in generally good agreement with the satellite observations, except for North America, Australia, and most remote oceans. Especially, over North America, the values of NICAM-simulated aerosol optical thickness were underestimated, possibly because of insufficient emissions from biomass burning and anthropogenic sources, or the lack of secondary organic aerosols formed from isoprene in forests (Henze and Seinfeld [2006]). Through the aerosol-cloud interaction, the cloud droplet effective radius generally decreases, as the aerosol optical thickness increases, which were generally captured by both NICAM and MODIS. The cloud droplet effective radii simulated by NICAM were smaller than those obtained from MODIS, probably because those retrieved from 2.2 μm (shown in Figure 8) correspond to those captured below the cloud top heights where the cloud droplet effective radii are larger than those at cloud top height (Nakajima et al. [2010]). Very recently, regional aerosol simulations that used the stretched grid system with a high spatial resolution of approximately 10 km were performed for the Kanto region of Japan under the MEXT/RECCA/SALSA project (SALSA [2014]; Goto et al. [2014]). These simulations were carried out at relatively low computational cost by using two supercomputers at the University of Tokyo (SR16000 and FX10).

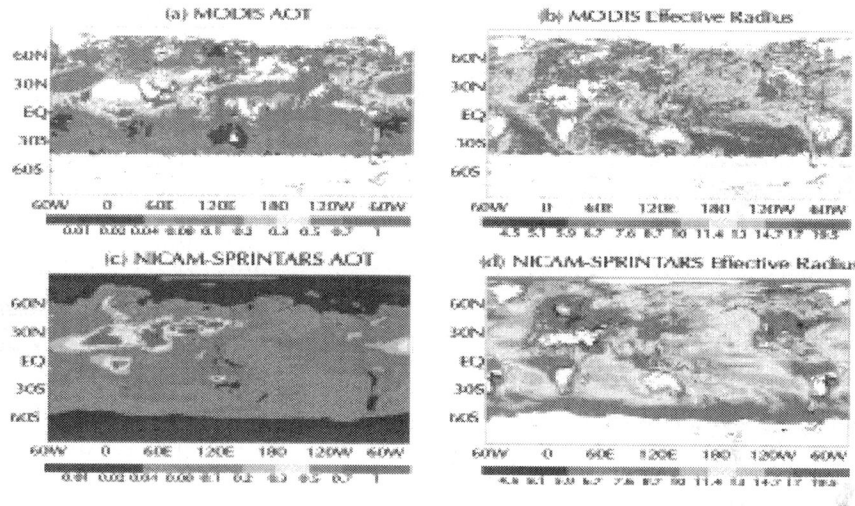

Figure 8: Distributions of aerosol optical thickness and cloud droplet effective radius from the NICAM-SPRINTARS simulations.Global geographical distributions of (a, c) aerosol optical thickness and (b, d) cloud droplet effective radius from (c, d) the NICAM-SPRINTARS simulations in comparison to those obtained from (a, b) the MODIS satellite observations for 1 to 8 July 2006 (cited from Suzuki et al. [2008]). The unit of cloud droplet effective radius is micrometers.

The CHASER atmospheric chemistry module has also been implemented into NICAM (SALSA[2014]). This module primary focuses on tropospheric chemistry while considering the chemical cycle of

Ox-NOx-HOx-CH$_4$-CO relation to the oxidation of volatile organic compounds (VOCs) and halogen chemistry in order to calculate the concentrations of 92 chemical species with 262 chemical reactions (58 photolytic, 183 kinetic, and 21 heterogeneous reactions) (Sudo et al.[2002a], [b]; Watanabe et al. [2011]). Using the CHASER module, the mixing ratios of atmospheric gases, which mainly include ozone, NOx, CO, SO$_2$, VOCs, and sulfate aerosols, can be predicted. The transport processes of the chemical tracers include emission, advection, diffusion, chemical reaction, wet deposition, and dry deposition. The emissions of lighting NOx and DMS are calculated by using the internal parameters of the model, whereas the emissions of the other gases are prescribed. The photolysis rates are calculated online by using the temperature and radiation fluxes computed in the radiative transfer module. Like the aerosol module, the meteorological field information is input from the outer module after the cloud microphysics calculations and to the chemistry module at each time step. In contrast, the calculated mass mixing ratios of the tracers in each grid cell within the chemistry module are returned back to the outer module. In addition, ozone, N$_2$O, and CH$_4$ are calculated online and used in the radiative transfer module to represent the atmospheric warming effects.

The distributions of the trace gases obtained by CHASER, which is currently coupled to MIROC at a low spatial resolution of approximately 300 km, are generally compared to the observations (Sudo et al. [2002b]; Sudo and Akimoto [2007]; Nagashima et al. [2010]) and to other chemistry models under international model comparison projects such as the Atmospheric Chemistry and Climate Model Intercomparison Project (ACCMIP) (e.g., Lamarque et al. [2013]). Under the MEXT/RECCA/SALSA project, the results of CHASER, along with the globally uniform and regionally stretched grid systems, were compared with the observations (SALSA [2014]). Furthermore, the chemical module has been currently expanded to a fully coupled aerosol-chemistry module by coupling CHASER with SPRINTARS, as developed in Watanabe et al. ([2011]) and used by MIROC.

The unified module, consisting of the fully coupled aerosol-chemistry, coupled to NICAM (hereinafter referred to as NICAM-Chem) can explicitly consider various interactions between aerosols and gases, as shown in Figure 9. For example, as an aspect of the aerosol simulation, the unified module can precisely calculate the sulfate formation with online oxidants such as hydroxyl radicals and ozone, whereas SPRINTARS uses assumed oxidants. In contrast, as an aspect of short-lived gases, calculations of the heterogeneous reactions onto the aerosol surface are expected to provide the greatest advantage when using the unified model. In addition to the sophisticated interactions between aerosols and gases, the chemistry of nitrate will be implemented into NICAM-Chem by using a thermodynamic equilibrium condition and secondary organic aerosols (SOAs). As a result, the new unified chemistry module is expected to provide better distributions of aerosols and gases, and, in the near future, we expect to be able to provide such distributions globally with a high spatial resolution of less than 10 km.

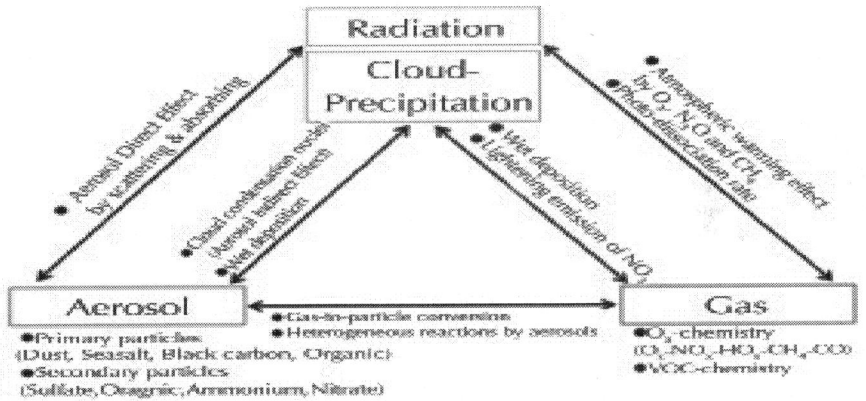

Figure 9: Relationship between the aerosol, gas, cloud, precipitation, and radiation fields for NICAM-Chem. Relationship between the aerosol, gas, cloud, precipitation, and radiation fields for the unified aerosol-chemistry module (NICAM-Chem). The aerosol and gas modules correspond to SPRINTARS and CHASER, respectively.

Land Model

We implemented two land surface schemes in NICAM in order to calculate the water and energy balance over land. The first scheme is a simple bucket land surface model (Kondo [1993]), and the other involves the minimal advanced treatments of surface interaction and runoff (MATSIRO) model (Takata et al. [2003]). When used with

NICAM, the simple bucket land surface model, which was first used by Miura et al. ([2007a]), has three vertical layers for predicting soil temperature and one layer for predicting the soil moisture.

The MATSIRO scheme is the same as that originally employed in MIROC (Hasumi and Emori[2004]). Kodama et al. ([2012]) were the first to use MATSIRO with NICAM. This scheme has five vertical layers for predicting soil temperature and soil moisture, as well as a snow scheme with thee vertical layers.

Even in a several-kilometer grid simulation, land use and land cover cannot be regarded as homogeneous in one grid cell if it is to be used to evaluate the land surface heterogeneities smaller than the horizontal grid scale. Accordingly, we are currently implementing a mosaic treatment of the land surface models in order to improve the accuracy of the water and energy fluxes at the land surface. We will also attempt to incorporate an urban canopy model in order to simulate the global urban energy balance.

Assimilation

Local Ensemble Transform Kalman Filter Applied to NICAM: NICAM-LETKF

The introduction of a data assimilation system into NICAM is a challenge because it might enable us to produce a high-resolution dataset with O (km) mesh size that can be used for improved better numerical simulations and research analysis. Data assimilation is an analysis technique that provides an accurate estimate of the atmospheric state by using observation data and model forecasts. It plays an important role in improving numerical weather prediction by providing an accurate initial condition. The ensemble Kalman filter (EnKF) is an advanced data assimilation method that uses flow-dependent forecast error statistics that are represented by the ensemble prediction (Evensen [1994]). For a linear system, the Kalman filter (KF) gives a minimum mean square error if the error is a Gaussian distribution (Kalman [1960]). However, it is difficult to apply the KF because the computational cost for an atmospheric model with a high degree of freedom (order of 10^7) is high. To resolve this problem, Evensen ([1994]) proposed the EnKF, which approximates the KF by using the forecast spread of the ensemble prediction. Furthermore, the local ensemble transform Kalman filter (LETKF) (Hunt et al. [2007]) was developed based on the ensemble transform Kalman filter (ETKF) (Bishop et al. [2001]) and used an algorithm that was designed for suitability with parallel computers by taking advantage of the independent local analyses of the local ensemble Kalman filter (LEKF) (Ott et al. [2004]). For details on the mathematical formulation and implementation of these approaches, refer to each reference.

Previous studies have applied the LETKF to regional and global atmospheric models and have shown promising results (Miyoshi and Aranami [2006]; Miyoshi and Yamane [2007]). Kondo and Tanaka ([2009]) applied the LETKF to NICAM in order to develop NICAM-LETKF, which was the first LETKF to be applied to a global non-hydrostatic model. They then investigated the feasibility and stability of NICAM-LETKF using a perfect model and determined that the system works appropriately for a realistic non-hydrostatic model (NICAM). However, the horizontal resolution used in their study was not very high (g-level 5 or $\overline{\Delta}$ = 220 km), where the model would behave hydrostatically. Additionally, it is unclear whether NICAM-LETKF will work stably when real observations are used. Therefore, a future task will be to investigate data assimilation with a high-resolution global non-hydrostatic regime using real observations. This would not only contribute to an improvement in weather prediction but also to the research on multi-scale interaction, by using the product of the assimilation. We are now developing a version of NICAM-LETKF that will be based on the LETKF code that was developed and continuously improved by Miyoshi (e.g., Miyoshi and Kunii [2011]) and are currently testing assimilation of precipitation data observed by satellites in order to obtain a fine-mesh precipitation database using the method proposed by Lien et al. ([2013]).

NICAM-Based Transport Model: NICAM-TM

An accurate estimation of the past and current CO_2 budget for the Earth's surface is needed for reliable predictions of carbon cycle changes under global warming conditions. Therefore, an inversion method coupled with a tracer transport model was used to estimate the spatial and temporal variations of CO_2 sources and sinks based on observations of CO_2 concentrations (e.g., Tans et al. [1990]). In an inversion analysis, we quantitatively estimated regional CO_2 fluxes by considering the *a priori* estimate and uncertainties of the CO_2 fluxes, which were derived from our present understanding of CO_2 source/sink mechanisms and uncertainties of observations (including model representative errors). This analysis is based on Bayesian theory and is similar to data assimilations for meteorological forecasts (Tarantola [2005]). However, in contrast to the observations of meteorological parameters, CO_2 concentrations are not sufficiently observed around the globe. Therefore, estimates of CO_2 sources and sinks are not well constrained by actual observations and are extremely sensitive to model transport errors (e.g., Gurney et al.[2002]).

In response to the need for an accurate transport model, Niwa ([2010]) developed the NICAM-based transport model (NICAM-TM). For transport simulations of CO_2, NICAM has an advantageous property: consistency with continuity (CWC) (Jöckel et al. [2001]; Gross et al. [2002]). Under the CWC condition, the mass conservation and Lagrangian conservation of a tracer are simultaneously guaranteed; that is, the volume integral of mass is conserved and a constant specific mass of a tracer is maintained along flow trajectories. The ability to achieve the CWC property is attributed to the fact that NICAM uses the finite volume method for its meteorological field integration and performs tracer transport on a common dynamical frame (Niwa et al. [2011a]). The tracer mass conservation is strictly required for analyses of the CO_2 budget, which is the main target of inversion studies. Furthermore, the Lagrangian conservation property is imperative for CO_2 because it is an abundantly existing tracer (the global average CO_2 concentration was about 390 ppm for 2013) and the concentration changes that are analyzed are very small (no more than a few tens of ppm).

NICAM-TM consists of online and offline transport models, as well as an adjoint model for the tracer transport. In the online model, tracer transport is calculated concomitantly but independently with the integration of the meteorological field. More specifically, the tracers other than those for water are passive. Compared to the online model, the offline model is computationally inexpensive because it only calculates the tracer transport by using meteorological data calculated by the online NICAM-TM and stored in a disk beforehand. The adjoint model was developed based on the offline model, which calculates the sensitivities to certain tracer variables backward in time. Because long-term and multiple simulations are required for the inverse calculation, NICAM-TM is currently run with a relatively low-resolution for CO_2 transport simulations; for the most part, a resolution of g-level 5 or 6 is used. However, after conducting intercomparison experiments (Law et al. [2008]; Patra et al. [2008]; Niwa et al. [2011b]), we found that the performances of CO_2 transport obtained are comparable to those of other models. Generally, the modeled winds are nudged towards the analyzed data (e.g., JMA Climate Data Assimilation System (JCDAS); Onogi et al. [2007]) in order to match the simulated CO_2 concentrations to the observed concentrations.

Niwa et al. ([2012]) performed an inversion analysis by using the synthesis inversion method (Enting [2002]) with NICAM-TM. This was the first inversion study in which aircraft data were extensively used to constrain the global and regional CO_2 budgets. The aircraft data were obtained from the Comprehensive Observation Network for Trace Gases by Airliner project (CONTRAIL) (Machida et al. [2008]; Matsueda et al. [2008]; Sawa et al. [2012]). By adding the CONTRAIL data (Figure 10a) to the conventionally used surface observation network, Niwa et al. ([2012]) successfully reduced uncertainties of the CO_2 flux estimates for Asia. Moreover, this study revealed the existence of a strong summer uptake by the biosphere in South Asia (Figure 10b), where the CO_2 observations from CONTRAIL in the middle to upper troposphere have a predominant constraint on the flux estimation. Because biosphere models have not shown such a strong summer uptake (Patra et al. [2013]), Niwa et al. ([2012]) suggested the existence of some unknown mechanisms of the CO_2 sources and sinks in this region. To further improve the CO_2 inversion, a four-dimensional (4D) variational method system is now being developed by using the adjoint model of NICAM-TM. Using this system, a large amount of CO_2 concentration data that have been recently available from regular aircraft observations and satellite observations (e.g., Greenhouse Gases Observing Satellite (GOSAT)) will be exploited to further constrain the estimates of CO_2 sources and sinks.

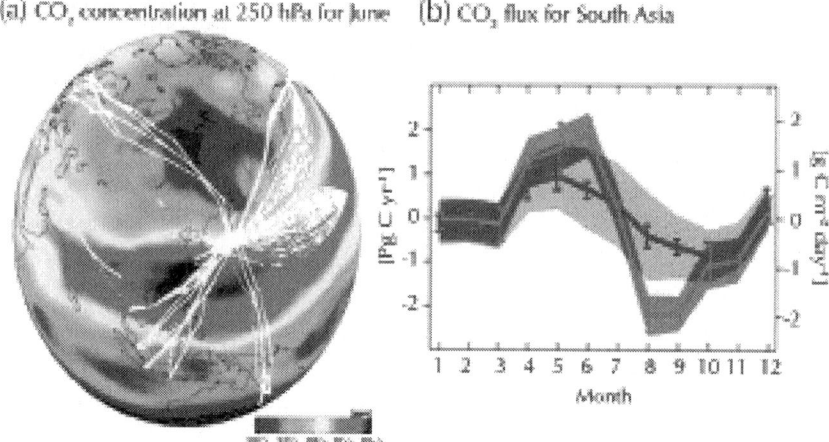

Figure 10: Simulated CO2concentration field and monthly variation of the CO2flux by NICAM-TM. (a) Mean CO_2 concentration field at 250 hPa for June 2007, as simulated by NICAM-TM, and the locations of the CONTRAIL measurements for 2007 (white dots). (b) Monthly variation of the CO_2 flux averaged for 2006 to 2008, estimated by the inversion analysis of Niwa et al. ([2012]) (this figure is slightly modified from Figure 5d in Niwa et al. [2012]). The blue and red lines are the CO_2 fluxes (positive values indicate sources and negative values indicate sinks) estimated by only using surface data and by using both surface and CONTRAIL data, respectively, and the colored shades and error bars denote the range of the estimated flux uncertainty and standard deviation for 2006 to 2008, respectively.

Models Related to NICAM

Stretch NICAM and Diamond NICAM

As described in 'Grid configuration and advection scheme' section, NICAM was originally designed for use with an icosahedral grid, which covers a sphere of the Earth quasi-uniformly and has a grid system based on 10 large diamonds. As extensions to the original configuration, NICAM can also be applied to a non-uniform resolution distribution grid or a regional grid by changing the positions of the grid points, the total number of diamonds, or the topology of their combination.

NICAM can also be adapted to simulations that target a particular area. Two approaches to performing a partially high-resolution simulation are typically used with NICAM. One includes the stretched icosahedral grid system developed by Tomita ([2008a]). In this approach, the icosahedral grid is stretched using the Schmidt transformation method and the grid size becomes gradually finer as grid points are concentrated onto the area of interest. In the stretch NICAM, the inhomogeneity of the horizontal grid sizes requires resolution-dependent parameters for physical schemes if the global domain is of interest. However, if a limited region is of interest, such as within regions with finer mesh sizes, resolution-dependent parameters are not required. The stretch NICAM is widely used for studies on tropical convective systems (Satoh et al. [2010]; Satoh and Kitao [2013]; Roh and Satoh [2014]), tropical cyclones (Yanase et al. [2010a], [b]; Satoh et al. [2013]; Arakane et al. [2014]), and MJOs (Nasuno [2013]).

The other approach is similar to the use of a regional model. Only 1 of the 10 diamond regions that make up the icosahedral grid is used for the simulation (hereinafter, the diamond NICAM). The simulation domain of the diamond NICAM consists of 1 diamond (2 triangles of an icosahedron), even though the original NICAM global domain consists of 10 diamonds (20 triangles). The diamond NICAM is used as a non-hydrostatic regional model. By using this approach, we can share the same code for both the global domain and limited-area models. This is advantageous for the maintenance of the source code because all components of NICAM are used in common in both models.

To enable the limited-area simulations using the diamond NICAM, we developed a nudging scheme in which the strength of the nudge depends on the distance from a certain point on the globe. Figure 11 shows the simulation domain of a regional climate simulation targeted on Japan. The lateral boundary of the simulation domain (outer edge of the gray circle in Figure 11) is forced by six-hourly reanalysis data. Verification of the diamond NICAM is still under investigation and the results of test simulations performed using the diamond NICAM were compared with the results of the high-resolution experiments using the stretch and global NICAM. The comparison results showed good agreement with those produced by the original global NICAM simulation. The temporal evolution of a convective system, such as the locations of triggering, development, and decay, was also well simulated during the integration.

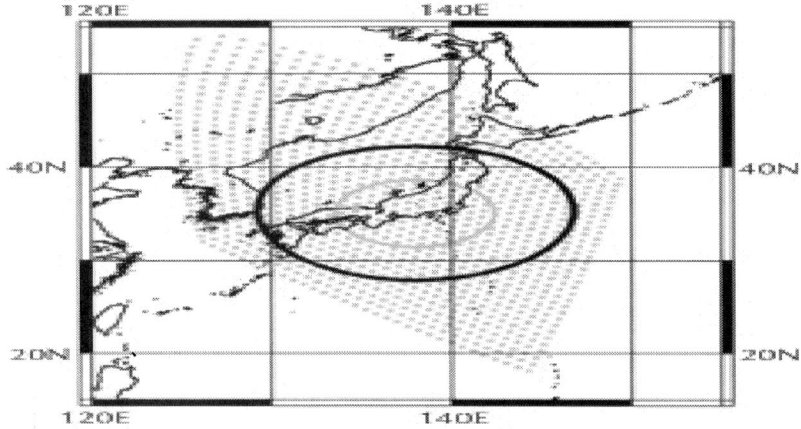

Figure 11: Simulation domain of the diamond NICAM targeted on Japan. Light blue dots indicate the grid points. Reanalysis nudging was applied to the outer side of the gray circle. This grid is based on g-level 5. Stretching parameter (in Equation 8, Tomita et al. [2008]) of grid transformation is 10. The minimum grid interval is about 59 km.

Plane NICAM

Although the stretch and diamond NICAM can be used for limited-area simulations, an *f*-plane system is useful when conducting some types of numerical simulations, such as idealized simulations on an *f*-plane and simulations with a narrow region in which the curvature of the Earth is not important. For this reason, an *f*-plane model (hereinafter, the plane NICAM) whose dynamical core is the same as the original NICAM was developed. To save maintenance costs, the plane NICAM used the same code as the global NICAM, which also makes knowledge gained by the plane NICAM immediately and directly applicable to the global NICAM.

In the plane NICAM, the shape of the calculation domain is a diamond composed of two regular triangles with a double periodic lateral boundary condition. Figure 12 shows the configuration of the calculation domain for the plane NICAM, as indicated by black bold lines. The outward fluxes across side AB are equivalent to the inward fluxes across side DC. The red, green, blue, and yellow areas inside the diamond area surrounded by the black bold lines correspond to the hatched areas of the same respective color. Therefore, if the double periodic boundary condition is applied to the diamond-shaped domain with side length L, the calculation domain is equivalent to the regular hexagonal domain with side length $l = L/\sqrt{3}$, as represented by the hatched area in Figure 12. Regular hexagonal geometry is better suited than square or regular triangle geometry in terms of isotropy.

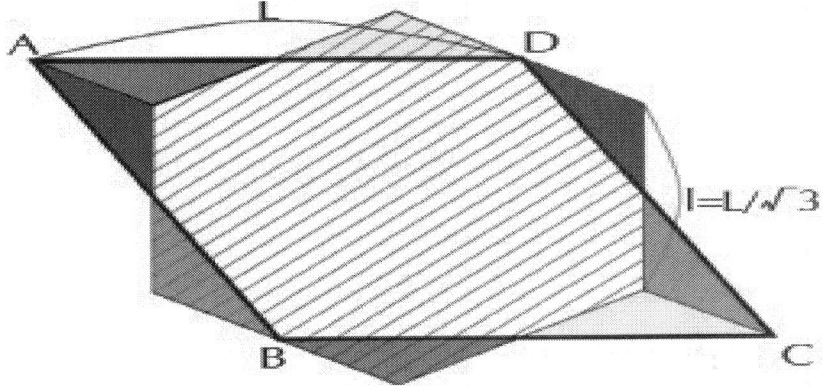

Figure 12: A schematic image of the calculation domain for the plane NICAM. A schematic image showing the relationship between the calculation domain and a unit cell for the plane NICAM. The diamond-shaped area (ABCD) surrounded by black bold lines indicates the calculation domain, and the hatched regular hexagonal area indicates the unit cell.

As an example of the use of the plane NICAM, we briefly show a result of an idealized tropical cyclone simulation on an f-plane (Ohno and Satoh [2014]). The model was initialized with an axisymmetric cyclonic vortex that had a maximum azimuthal wind of 12 m s^{-1} as formulated by Rotunno and Emanuel ([1987]). The initial thermodynamic structure of the unperturbed model atmosphere is defined by moist adiabatic lapse rate up to 17 km and a constant temperature above this height. The length of the sides of the numerical domain is $L = 3,000$ km, and a horizontal resolution of 2.7 km is used for the entire domain. We used 50 vertical levels up to 45 km where the intervals of the vertical levels become smaller in the boundary layer. The SST is fixed at 31°C and an f-plane with a constant Coriolis parameter at 18 N is assumed. After an initial shock, the minimum sea level pressure continued to drop for 140 h and a quasi-steady stage was achieved. The cyclone reached a minimum central surface pressure of 931 hPa at about 160 h. A plan view of the distribution of the hourly precipitation amounts after 128 h of the time integration is shown in Figure 13, which indicates that the tropical cyclone has an eye, an eyewall, inner core rainbands, and distant rainbands. The array of rainbands and the eyewall are typical features of a tropical cyclone (Houze [2010]).

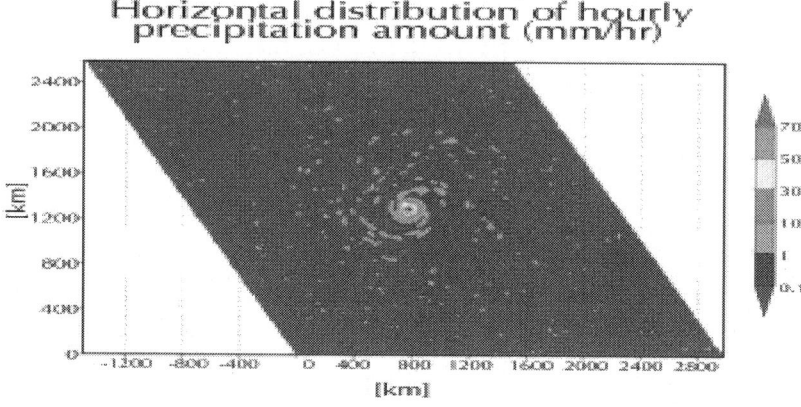

Figure 13: A horizontal distribution of precipitation for a tropical cyclone simulation by the plane NICAM. A horizontal distribution of hourly precipitation amount by the plane NICAM after a time integration of 128 h for a 2.7 km mesh tropical cyclone simulation.

Further Modified Grids

In addition to the original icosahedral grid, alternative spherical grids denoted by extended triangular meshes (XTMS) (Iga [2014]a) are also implemented on NICAM. The mesh topologies of these grids are different from the topology of an icosahedral grid and are generated by three processes: grid relaxation via spring dynamics, transformation via an analytical function, and the Schmidt transformation. By changing mesh topologies and transformation functions, it is possible to obtain a grid with non-uniform distributions of resolution on a sphere, which is applicable to various situations. For example, the resolution can be increased at a particular region or along a tropical zone as shown by Figure 14a,b. The grid structure shown by Figure 14a looks similar to that of the stretch NICAM, but the variability of the horizontal mesh spacing is smaller and the variability of the resolution is smoother than that of the stretch NICAM. The resolution is approximately proportional to the combination of map factors of two polar stereographic projections. As for the grid structure shown by Figure 14b, resolution is enhanced along the equatorial belt and is approximately proportional to the combination of map factors of two Lambert conformal conic projections and one Mercator projection (hereinafter, LML grid). The LML grid is more useful for studying the multi-scale structure of convective systems in the tropics. The performance of the LML grid was examined by aqua planet experiments (Iga [2014]b).

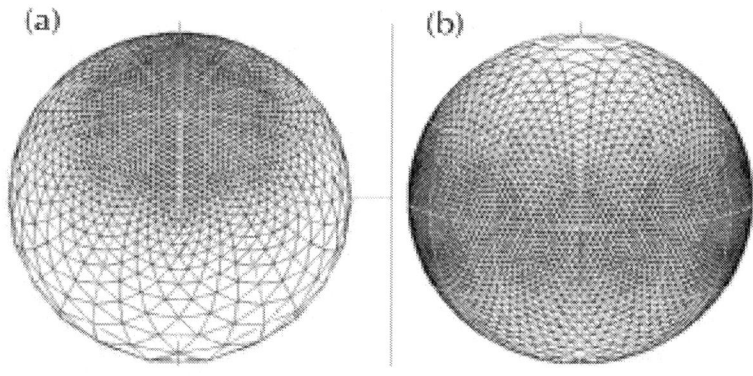

Figure 14: Grid configurations of the regionally enhanced grid and the tropically enhanced grid. Grid configurations of (a) the regionally enhanced grid and (b) the tropically enhanced grid (after Iga [2014]a).

Coupler

Currently, high-resolution NICAM is mainly used for short-term simulations such as for tropical cyclones and intra-seasonal variability. Since deep sea circulations are not important for these simulations, a simple mixed-layer ocean model is implemented in NICAM. However, in order to cover wider areas of research, such as climate projection, coupling with an ocean model is necessary when NICAM is used as an atmospheric component of an Earth system model. To couple an atmosphere model and an ocean model, we used a coupler called Jcup (Figure 15). Because Jcup connects models with different grid systems, it can be used as a grid transformation tool. Thus, in addition to ocean coupling, Jcup is used as an I/O tool of NICAM by converting the output data from the icosahedral grid to the lat-lon grid which is generally used for analysis. In this section, the Jcup coupling library that was used as a core of the coupled model system will be introduced first, after which the ocean and I/O coupling will be described.

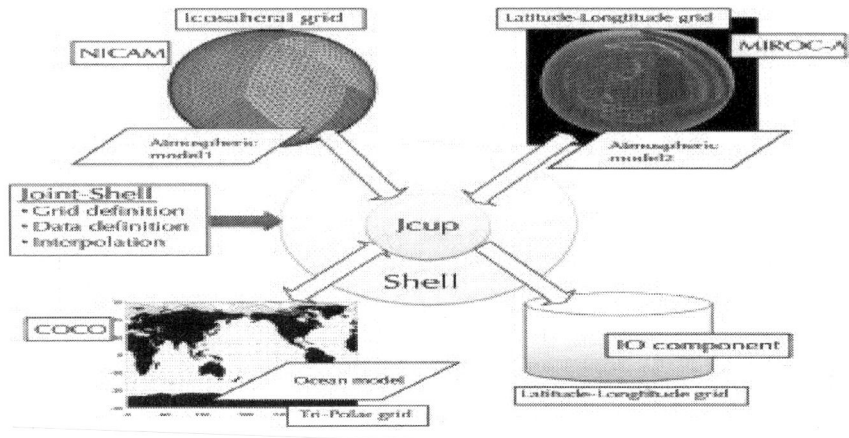

Figure 15: A schematic of the coupling system.

The Jcup coupling library is a collaborative work of JMA/MRI, JAMSTEC, and the Research Organization for Information Science and Technology (RIST) (Arakawa and Yoshimura [2009]; Arakawa et al. [2011]), that is freely available as an open source program athttps://sites. google.com/a/rist.jp/jcup/ *webcite*. Jcup inherited the development experience and design concept of Scup, which was developed by Yoshimura and Yukimoto ([2008]). The most remarkable feature of Jcup is its wide applicability. Generally speaking, the applicability range of a coupler is restricted by the specific grid structures and interpolation schemes it supports. For example, OASIS version 4 supported logically rectangular (i.e., 2D structured) or reduced Gaussian grids, along with bilinear, trilinear, bicubic, nearest-neighbor, and 2D conservative interpolations (Redler et al. [2009]). This indicates that the models that utilize other grid systems, such as non-structured or substructured grids, cannot be coupled.

The spectral method has been widely used for global atmospheric models. However, this method is not suitable for massively parallel processor (MPP) supercomputers because all-to-all communication is required. This is one reason why substructured grid models such as NICAM were developed energetically by the research organizations of numerous countries. Therefore, it is also necessary to develop a coupler that can handle such models. Jcup is specifically designed to couple different grid systems, so users can (and must) implement their own interpolation code and prepare a 'mapping table' in advance. See Arakawa et al. ([2011]) for details.

Next, we will describe NICAM and ocean model coupling. Currently, the Center for Climate System Research (CCSR) Ocean Component Model (COCO) (Hasumi [2000], [2006]) is used for the ocean model coupled with NICAM. In the coupler of the NICAM-COCO coupled model, Jcup is used as a core library and is implemented via a package called Joint-shell, which contains definition of the grid structures, exchange data, and the interpolation code (Figure 15). The interpolation code is based on the first-order conservative remapping scheme in Jones ([1999]). The 'mapping table' that must be set up as external files for beforehand defines grid-to-grid relations and Jcup's weight for grid systems of coupled models. Furthermore, we also implement a utility program to calculate such table files as a part of the Joint-shell. Note that COCO adopts a tri-polar grid, in which the northern polar region grid points do not follow latitude-longitude lines. As a result, coupling NICAM and COCO requires coupling two substructured grid models, which is why Jcup is used as a core of the coupler.

To couple NICAM, a new module has been implemented to replace the mixed-layer ocean module that has already implemented. Most of the other parts, such as the dynamical core, cloud microphysics, and land models, are untouched. The physical quantities exchanged between NICAM and COCO are listed in Table 3. The most significant issue was the inconsistency of ocean definition points between the two models. More specifically, some points that are recognized as ocean points by NICAM are recognized as land points by COCO. At such grid points, no data are provided to NICAM. Inconsistencies such as this occur if the land-ocean distribution has been set independently between the two models. In the future, we plan to improve pre-process utilities in order to create a NICAM land-ocean map that will maintain consistency with the COCO map. However, to avoid this problem in the interim, we applied the mixed-layer ocean module, which is pre-implemented in NICAM to such grid points.

Table 3: Physical quantities exchanged between NICAM and COCO through the coupler

NICAM to COCO	COCO to NICAM
Wind stress to ocean (eastward)	Sea surface temperature
Wind stress to ocean (northward)	Sea ice thickness
Wind stress to sea ice (eastward)	Ice covered ratio
Wind stress to sea ice (northward)	Snow amount on sea ice
Energy flux from atmosphere to ocean	Temperature of sea ice
Energy flux from atmosphere to sea ice	
Energy flux from sea ice to ocean	
Radiation flux (short wave)	
Sublimation from sea ice	
Evaporation	
Precipitation	
Snowfall	
Runoff	

Satoh et al.

Satoh et al. Progress in Earth and Planetary Science 2014 1:18, doi:10.1186/s40645-014-0018-1

The icosahedral grid employed by NICAM cannot be used to analyze the results. For example, the calculation of the zonal mean values is not straightforward and the visualization tools assume a lat-lon grid in many cases. Because of this, we generally use a conversion tool called 'ico2ll' to move data from the icosahedral grid to the lat-lon grid for post-processing of simulations. However, in many cases, running ico2ll as a post-process results in a bottle neck because it requires numerous I/O transfers and cannot parallelize itself since it is necessary to create a single file for the lat-lon grid of each variable. Jcup allows NICAM-I/O coupling, which leads to an efficient I/O conversion, to be successfully achieved. More specifically, the I/O module converts the icosahedral grid to the lat-lon grid and is executed in parallel with NICAM. The implemented conversion schemes include a bi-linear interpolation, a control volume weighted average, and the nearest-neighbor method. A bi-linear interpolation is originally implemented method in the stand-alone version of ico2ll', and this method has first-order accuracy. The control volume weighted average scheme is the same with the one used in NICAM-COCO coupling and also has first-order accuracy, but the global averaged fluxes are conserved. The nearest-neighbor method has less accuracy and conservation properties than the other two methods. This method was implemented due to the demands from NICAM users and is used to provide a quick look at the raw values by using the general latitude-longitude graphic tools, such as the Grid Analysis and Display System (GrADS), without an averaging process. These schemes can be set for each output data type and can be switched by altering the configuration file.

Figure 15 is a schematic of the coupling system described above. The coupling system is designed so that NICAM automatically detects the coupling pattern at runtime without taking any other configuration. For example, when COCO is executed in parallel with NICAM, subroutines for NICAM-COCO coupling are used, and if not, subroutines of the mixed-layer ocean model are called. I/O is also the same as the case of COCO, in which NICAM automatically sends output data to I/O only when the I/O component is executed.

CONCLUSIONS

This paper reviewed recent activities of the global non-hydrostatic model, NICAM, and described its development and the details of each component of the dynamics and physics, the strategy of assimilation, and the related models. In the 'Introduction' section, a review of other global non-hydrostatic models, scientific overviews of NICAM, and the current design/structure and a future timeframe of NICAM were described. In the 'Dynamics' section, horizontal grid issues and vertical resolution issues were discussed. In the 'Physics' section, components of cloud microphysics, subgrid-scale turbulence, radiation, aerosol and chemistry, and land surface were described. As for assimilation, NICAM-LETKF was presented, and NICAM-TM was introduced to describe the assimilation of the transport model. The 'Models related to NICAM' section showed a variety of ways that NICAM can be used, specifically, stretch NICAM, diamond NICAM, plane NICAM, the tropically enhanced grid model, and coupling with the ocean model or the I/O module.

A number of comprehensive NICAM-related review articles have been published (Satoh et al.[2008]; Satoh [2013]) since the 3.5 km mesh global non-hydrostatic experiments were conducted by Tomita et al. ([2005]) and Miura et al. ([2007b]). However, these were primarily limited to descriptions of the governing equations, and because many of the model components continue to evolve and develop as the entire package expands, we were requested to provide a comprehensive overview of NICAM activities and future projections. NICAM has been a pioneering global non-hydrostatic model for almost a decade, and now that other global non-hydrostatic models are emerging, we believe that our experiences and considerations will be useful for other high-resolution modeling groups. Mutually beneficial information exchanges between different modeling groups will be required for future fruitful studies during the forthcoming era of global non-hydrostatic modeling.

AUTHORS' CONTRIBUTIONS

MS and HT coordinated the structure of the paper and wrote the 'Introduction' and 'Conclusions' sections. HY wrote the 'Design, structure, development, and timeline' and 'Radiation' sections. In the 'Dynamics' section, HM wrote the 'Grid configuration and advection scheme' section, while CK and MS wrote the 'Vertical resolution issues' section. In the 'Physics' section, TS wrote the 'Cloud microphysics schemes' section, while AN wrote the 'Subgrid-scale turbulence of the planetary boundary layer' section using experimental data simulated by YY. DG wrote the 'Aerosol and chemistry modules' section, and MH wrote the 'Land model' section. In the 'Assimilation' section, MSw and TM wrote the 'Local ensemble transform Kalman filter applied to NICAM: NICAM-LETKF' section, while YN wrote the 'NICAM-based transport model: NICAM-TM' section. In the 'Models related to NICAM' section, MH wrote the 'Stretch NICAM and diamond NICAM' section, while TO wrote the 'Plane NICAM' section. The 'Further modified grids' section was written by SI, and the 'Coupler' section was written by TA, TI, and HK. All authors read and approved the final manuscript.

ACKNOWLEDGEMENTS

This research used the computational resources of the High Performance Computing Infrastructure (HPCI) system provided by the Information Technology Center of The University of Tokyo, through the HPCI System Research Project (Project ID:hp120190). The K computer at AICS was used under the supported by Strategic Programs for Innovative Research (SPIRE) Field 3 (Projection of Planet Earth Variations for Mitigating Natural Disasters), which is funded by MEXT (ID: hp120279 and hp130010). Earth Simulator at JAMSTEC and FX10 at The University of Tokyo were also used. The development of coupler is supported by Japan Science and Technology Agency/Core Research for Evolutional Science and Technology (JST/CREST), 'ppOpen-HPC: Open Source Infrastructure for Development and Execution of Large-Scale Scientific Applications on Post-Peta-Scale Supercomputers with Automatic Tuning (AT)'. The development of the diamond NICAM and NICAM-Chem is supported by MEXT/RECCA/SALSA.

REFERENCES

1. Adcroft A, Hill C, Marshall J (1997) Representation of topography by shaved cells in a height coordinate ocean model. Mon Wea Rev 125:2293-2315

2. Albrecht BA (1989) Aerosols, cloud microphysics, and fractional cloudiness. Science 245:1227-1230

3. Arakane T, Satoh M, Yanase W (2014) The excitation of the deep convection to the north of tropical storm Bebinca (2006). J Meteor Soc Japan 92:141-161

4. Arakawa A (2004) The cumulus parameterization problem: past, present, and future. J Clim 17:2493-2525

5. Arakawa A, Lamb VR (1977) Computational design of the basic dynamical processes of the UCLA general circulation model. Methods Comput Phys 17:173-265

6. Arakawa T, Yoshimura H (2009) Performance evaluation of a coupling software for climate modeling (in Japanese). J Inform Process Japan 2:95-110

7. Arakawa T, Yoshimra H, Saito F, Ogochi K (2011) Data exchange algorithm and software design of KAKUSHIN coupler Jcup. Procedia Comp Sci 4:1516-1525

8. Berger AL (1978) Long-term variations of daily insolation and quaternary climatic changes. J Atmos Sci 35:2362-2367

9. Berry EX, Reinhardt RL (1974) An analysis of cloud drop growth by collection: part II. Single initial distribution. J Atmos Sci 31:1825-1831

10. Bishop CH, Etherton B, Majumdar SJ (2001) Adaptive sampling with the ensemble transform Kalman filter. Part I: theoretical aspects. Mon Wea Rev 129:420-436

11. Blackburn M, Williamson DL, Nakajima K, Ohfuchi W, Takahashi YO, Hayashi Y-Y, Nakamura H, Ishiwatari M, McGregor J, Borth H, Wirth V, Frank H, Bechtold P, Wedi NP, Tomita H, Satoh M, Zhao M, Held IM, Suarez MJ, Lee M-I, Watanabe M, Kimoto M, Liu Y, Wang Z, Molod A, Rajendran K, Kitoh A, Stratton R (2013) The Aqua-Planet Experiment (APE): Control SST simulation. J Meteor Soc Japan 91A:17-56

12. Bodas-Salcedo A, Webb MJ, Bony S, Chepfer H, Dufresne JL, Klein SA, Marchand R, Haynes JN, Pincus R, John VO (2011) COSP: satellite simulation software for model assessment. Bull Am Meteorol Soc 91:1023-1043

13. Bonaventura L (2000) A semi-implicit semi-Lagrangian scheme using the height coordinate for a nonhydrostatic and fully elastic model of atmospheric flows. J Comp Phys 158:186-213

14. Bryan GH, Wyngaard JC, Fritsch JM (2003) Resolution requirements for the simulation of deep moist convection. Mon Wea Rev 131:2394-2416

15. Cheong H-B (2006) A dynamical core with double Fourier series: comparison with the spherical harmonics method. Mon Wea Rev 134:1299-1315

16. Collins WD, Satoh M (2009) Simulating global clouds, past, present, and future. In: Heintzenberg J, Charlson RJ (eds) Clouds in the perturbed climate system: their relationship to energy balance, atmospheric dynamics, and precipitation, Struengmann Forum Report, vol. 2, The MIT Press, Cambridge. pp 469-486

17. Côté J, Gravel S, Méthot A, Patoine A (1998) The operational CMC-MRB global environmental multiscale (GEM) model. Part I: design considerations and formulation. Mon Wea Rev 126:1373-1395

18. Cullen MJP, Davies T, Mawson MH, James JA, Coutler SC, Malcolm A, et al. (1997) An Overview of Numerical Methods for the Next Generation U. K. NWP and Climate Model. In: Lin CA (ed) Numerical Methods in Atmospheric and Oceanic Modelling, The Andrew J. Robert Memorial Volume, NRC Research Press, Ottawa. pp 425-444

19. Dai T, Goto D, Schutgens NAJ, Dong X, Shi G, Nakajima T (2014) Simulated aerosol key optical properties over global scale using an aerosol transport coupled with a new type of dynamic core. Atmos Environ 82:71-82

20. Dai T, Schutgens NAJ, Goto D, Shi G, Nakajima T (2014b) Improvement of aerosol optical properties modeling over Eastern Asia with MODIS AOD assimilation in a global non-hydrostatic icosahedral aerosol transport model. Environ Pollut doi:10.1016/j.envpol.2014.06.021

21. Davies T, Cullen MJP, Malcolm AJ, Mawson MH, Staniforth A, White AA, Wood N (2005) A new dynamical core for the Met Office's global and regional modelling of the atmosphere. Quart J Roy Meteor Soc 131:1759-1782

22. Dirmeyer PA, Cash BA, Kinter JL III, Jung T, Marx L, Satoh M, Stan C, Tomita H, Towers P, Wedi N, Achuthavarier D, Adams JM, Altshuler EL, Huang B, Jin EK, Manganello J (2012) Simulating the diurnal cycle of rainfall in global climate models: resolution versus parameterization. Clim Dyn 39:399-418

23. Emanuel K, Oouchi K, Satoh M, Tomita H, Yamada Y (2010) Comparison of explicitly simulated and downscaled tropical cyclone activity in a high-resolution global climate model. J Adv Model Earth Syst 2:9 doi:10.3894/JAMES.2010.2.9

24. Enting IG (2002) Inverse problems in atmospheric constituent transport. Cambridge University Press, Cambridge.

25. Evensen G (1994) Sequential data assimilation with a nonlinear quasi-geostrophic model using Monte-Carlo methods to forecast error statistics. J Geophys Res 99:10143-10162

26. Feingold G, Walko RL, Stevens B, Cotton WR (1998) Simulations of marine stratocumulus using a new microphysical parameterization scheme. Atmos Res 47–48:505-528

27. Fu X, Wang B (2004) Differences of boreal summer intraseasonal oscillations simulated in an atmosphere–ocean coupled model and an atmosphere-only model. J Clim 17:1263-1271

28. Fudeyasu H, Wang Y, Satoh M, Nasuno T, Miura H, Yanase W (2008) The global cloud-system-resolving model NICAM successfully simulated the lifecycles of two real tropical cyclones. Geophys Res Lett 35:L22808 doi:10.1029/2008GL0360033

29. Fudeyasu H, Wang Y, Satoh M, Nasuno T, Miura H, Yanase W (2010) Multiscale interactions in the lifecycle of a tropical cyclone simulated in a global cloud-system-resolving model. Part I: large scale aspects. Mon Wea Rev 138:4285-4304

30. Fudeyasu H, Wang Y, Satoh M, Nasuno T, Miura H, Yanase W (2010) Multiscale interactions in the lifecycle of a tropical cyclone simulated in a global cloud-system-resolving model. Part II: mesoscale and storm-scale processes. Mon Wea Rev 137:3254-3268

31. Fujita M, Yoneyama K, Mori S, Nasuno T, Satoh M (2011) Diurnal convection peaks over the eastern Indian Ocean off Sumatra during different MJO phases. J Meteor Soc Japan 89A:317-330

32. Gal-Chen T, Somerville CJ (1975) On the use of a coordinate transformation for the solution of the Navier-Stokes equations. J Comp Phys 17:209-228

33. Gallus WA Jr, Klemp JB (2000) Behaviour of flow over step orography. Mon Wea Rev 128:1153-1164

34. Goto D (2014) Modeling of black carbon in Asia using a global-to-regional seamless aerosol-transport model. Environ Pollut doi:10.1016/j.envpol.2014.06.006

35. Goto D, Nakajima T, Takemura T, Sudo K (2011) A study of uncertainties in the sulfate distribution and its radiative forcing associated with sulfur chemistry in a global aerosol model. Atmos Chem Phys 11:10889-10910

36. Goto D, Takemura T, Nakajima T, Badarinath KVS (2011) Global aerosol model-derived black carbon concentration and single scattering albedo over Indian region and its comparison with ground observations. Atmos Environ 45:3277-3285

37. Goto D, Kanazawa S, Nakajima T, Takemura T (2012) Evaluation of a relationship between aerosols and surface downward shortwave flux through an integrative analysis of modeling and observation. Atmos Environ 49:294-301

38. Goto D, Dai T, Satoh M, Tomita H, Uchida J, Misawa S, Inoue T, Tsuruta H, Ueda K, Ng CFS, Takami A, Sugimoto N, Shimizu A, Ohara T, Nakajima T (2014) Application of a global nonhydrostatic model with a stretched-grid system to regional aerosol simulations around Japan. Geosci Model Dev Discuss 7:131-179

39. Grabowski WW (1998) Toward cloud resolving modeling of large-scale tropical circulations: a simple cloud microphysics parameterization. J Atmos Sci 55:3283-3298

40. Gross ES, Bonaventura L, Rosatti G (2002) Consistency with continuity in conservative advection schemes for free-surface models. Int J Numer Meth Fluids 38:307-327

41. Gurney KR, Law RM, Denning AS, Rayner PJ, Baker D, Bousquet P, Bruhwiler L, Chen YH, Ciais P, Fan S, Fung IY, Gloor M, Heimann M, Higuchi K, John J, Maki T, Maksyutov S, Masarie L, Peylin P, Prather M, Pak BC, Randerson J, Sarmiento J, Taguchi S, Takahashi T, Yuen CW (2002) Towards robust estimates of CO_2 sources and sinks using atmospheric transport models. Nature 415:626-630

42. Ham S-H, Sohn B-J, Kato S, Satoh M (2013) Vertical inhomogeneity of ice cloud layers from CloudSat and CALIPSO measurements and comparison to NICAM simulations. J Geophys Res 118:9930-9947

43. Hansen JE, Travis LD (1974) Light scattering in planetary atmosphere. Space Sci Rev 16:527-610

44. Hashino T, Satoh M, Hagihara Y, Kubota T, Matsui T, Nasuno T, Okamoto H (2013) Evaluating global cloud distribution and microphysics from the NICAM against CloudSat and CALIPSO. J Geophys Res 118:7273-7292

45. Hasumi H (2000) CCSR Ocean Component Model (COCO). CCSR Rep 13. The University of Tokyo, Chiba, Japan.

46. Hasumi H (2006) CCSR Ocean Component Model (COCO) version 4.0. CCSR Rep 25. The University of Tokyo, Chiba, Japan.

47.   [http://ccsr.aori.u-tokyo.ac.jp/~hasumi/miroc_description.pdf] *webcite* Hasumi H, Emori S (2004) K-1 coupled model (MIROC) description. K-1 Tech Rep. p 34, . Accessed at 29 Sep 2014

48.   Hayashi YY, Sumi A (1986) The 30–40 day oscillations simulated in an aqua-planet model. J Meteorol Soc Japan 64:451-467

49.   Heikes R, Randall DA (1995) Numerical integration of the shallow-water equations on a twisted icosahedral grid. Part I: basic design and results of tests. Mon Wea Rev 123:1862-1880

50.   Heikes R, Randall DA, Konor CS (2013) Optimized icosahedral grids: performance of finite-difference operators and multigrid solver. Mon Wea Rev 141:445-4469

51.   Henze DK, Seinfeld JH (2006) Global secondary organic aerosol from isoprene oxidation. Geophys Res Lett 33:L09812 doi:10.1029/2006GL025976

52.   Holloway CE, Woolnough SJ, Lister GMS (2013) The effects of explicit versus parameterized convection on the MJO in a large-domain high-resolution tropical case study. Part I: characterization of large-scale organization and propagation. J Atmos Sci 70:1342-1369

53.   Hong SY, Dudhia J, Chen SH (2004) A revised approach to ice microphysical processes for the bulk parameterization of clouds and precipitation. Mon Wea Rev 132:103-120

54.   Houze RA (2010) Clouds in tropical cyclones. Mon Wea Rev 138:293-344

55.   Hunt BR, Kostelich EJ, Szunyogh I (2007) Efficient data assimilation for spatiotemporal chaos: a local ensemble transform Kalman filter. Physica D 230:112-126

56.   Iga S (2014a) Smooth, seamless and structured grid generation with flexibility in resolution distribution on a sphere based on conformal mapping and spring dynamics method. J Compt Phys submitted

57.   Iga S (2014b) Aqua-planet experiment on an AGCM with tropics-enhanced grid. SOLA submitted

58.   Iga S, Tomita H (2014) Improved smoothness and homogeneity of icosahedral grids using the spring dynamics method. J Comp Phys 258:208-226

59. Iga S, Tomita H, Tsushima Y, Satoh M (2007) Climatology of a nonhydrostatic global model with explicit cloud processes. Geophys Res Lett 34:L22814 doi:10.1029/2007GL031048

60. Iga S, Tomita H, Satoh M, Goto K (2007) Mountain-wave-like spurious waves due to inconsistency of horizontal and vertical resolution associated with cold fronts. Mon Wea Rev 135:2629-2641

61. Iga S, Tomita H, Tsushima Y, Satoh M (2011) Sensitivity of upper tropospheric ice clouds and their impacts on the Hadley circulation using a global cloud-system resolving model. J Climate 24:2666-2679

62. Inoue T, Satoh M, Miura H, Mapes B (2008) Characteristics of cloud size of deep convection simulated by a global cloud resolving model. J Meteor Soc Japan 86A:1-15

63. Inoue T, Satoh M, Hagihara Y, Miura H, Schmetz J (2010) Comparison of high-level clouds represented in a global cloud system-resolving model with CALIPSO/CloudSat and geostationary satellite observations. J Geophys Res 115:D00H22 doi:10.1029/2009JD012371

64. Iwasaki T, Yamada S, Tada K (1989) A parameterization scheme of orographic gravity wave drag with two different vertical partitionings. Part I: impacts on medium-range forecasts. J Meteor Soc Japan 67:11-27

65. Jöckel P, von Kuhlmann R, Lawrence M, Steil B, Brenninkmeijer C, Crutzen P, Rasch P, Eaton B (2001) On a fundamental problem in implementing flux-form advection schemes for tracer transport in 3-dimensional general circulation and chemistry transport models. Quart J Roy Meteor Soc 127:1035-1052

66. Jones PW (1999) First- and second-order conservative remapping schemes for grids in spherical coordinates. Mon Wea Rev 127:2204-2210

67. Jung T, Miller MJ, Palmer TN, Towers P, Wedi N, Achuthavarier D, Adams JM, Altshuler EL, Cash BA, Kinter JL III (2012) High-resolution global climate simulations with the ECMWF model in Project Athena: experimental design, model climate, and seasonal forecast skill. J Climate 25:3155-3172

68. Kalman RE (1960) A new approach to linear filtering and prediction problems. J Basic Eng Trans ASME 82:35-45

69. Kato S, Rose FG, SunMack S, Miller WF, Chen Y, Rutan DA, Stephens GL, Loeb NG, Minnis P, Wielicki BA, Winker DA, Charlock TP, Stackhouse PW Jr, Xu KM, Collins WD (2011) Improvements of top-of-atmosphere and surface irradiance computations with CALIPSO-, CloudSat-, and MODIS-derived cloud and aerosol properties. J Geophys Res 116:D19209 doi:10.1029/2011JD016050

70. Kessler E (1969) On the distribution and continuity of water substance in atmospheric circulations. Meteorologial Monograph. Amer Meteor Soc 32:1-84

71. Khain A, Ovtchinnikov M, Pinsky M, Pokrovsky A, Krugliak H (2000) Notes on the state-of-the-art numerical modeling of cloud microphysics. Atmos Res 55:159-224

72. Khain A, BenMoshe N, Pokrovsky A (2008) Factors determining the impact of aerosols on surface precipitation from clouds: an attempt at classification. J Atmos Sci 65:1721-1748

73. Khairoutdinov M, Kogan Y (2000) A new cloud physics parameterization in a large-eddy simulation model of marine stratocumulus. Mon Wea Rev 128:229-243

74. Khairoutdinov M, Krueger SK, Moeng CH, Bogenschutz PA, Randall DA (2009) Large-eddy simulation of maritime deep tropical convection. J Adv Model Earth Systems 1:15 doi:10.3894/JAMES.2009.1.15

75. Kikuchi K, Wang B (2010) Formation of tropical cyclones in the northern Indian Ocean associated with two types of tropical intraseasonal oscillation modes. J Meteor Soc Japan 88:475-496

76. Kinne S, Schulz M, Textor C, Guibert S, Balkanski Y, Bauer SE, Berntsen T, Berglen TF, Boucher O, Chin M, Collins W, Dentener F, Diehl T, Easter R, Feichter J, Fillmore D, Ghan S, Ginoux P, Gong S, Grini A, Hendricks J, Herzog M, Horowitz L, Isaksen I, Iversen T, Kirkevag A, Kloster S, Koch D, Kristjansson JE, Krol M, et al. (2006) An AeroCom initial assessment - optical properties in aerosol component modules of global models. Atmos Chem Phys 6:1815-1834

77. Kinter JL III, Cash B, Achuthavarier D, Adams J, Altshuler E, Dirmeyer P, Doty B, Huang B, Marx L, Manganello J, Stan C, Wakefield T, Jin E, Palmer T, Hamrud M, Jung T, Miller M, Towers P, Wedi N, Satoh M, Tomita H, Kodama C, Nasuno T, Oouchi K, Yamada Y, Taniguchi H, Andrews P, Baer T, Ezell M, Halloy C, et al. (2013) Revolutionizing climate modeling - Project Athena: a multi-institutional, international collaboration. Bull Am Meteor Soc 94:231-245

78. Klein SA, Jakob C (1999) Validation and sensitivities of frontal clouds simulated by the ECMWF model. Mon Wea Rev 127:2514-2531

79. Klemp JB (2011) A terrain-following coordinate with smoothed coordinate surfaces. Mon Wea Rev 139:2163-2169

80. Klemp JB, Wilhemson RB (1978) The simulation of three-dimensional convective storm dynamics. J Atmos Sci 35:1070-1096

81. Kodama C, Noda AT, Satoh M (2012) An assessment of the cloud signals simulated by NICAM using ISCCP, CALIPSO, and CloudSat satellite simulators. J Geophys Res 117:D12210 doi:10.1029/2011JD017317

82. Kodama C, Iga S, Satoh M (2014) Impact of the sea surface temperature rise on storm-track clouds in global non-hydrostatic aqua-planet simulations. Geophys Res Lett 41:3545-3552 doi:10.1002/2014GL059972

83. Kodama C, Terai M, Noda AT, Yamada Y, Satoh M, Seiki T, Iga S, Yashiro H, Tomita H, Minami K (2014) Scalable rank-mapping algorithm for an icosahedral grid system on the massive parallel computer with a 3-D torus network. Parallel Comput 40:362-373

84. Kondo J (1993) A new bucket model for predicting water content in the surface model. J Japan Soc Hydrol and Water Resour 6:344-349 (in Japanese with English abstract)

85. Kondo K, Tanaka HL (2009) Applying the local ensemble transform Kalman filter to the Nonhydrostatic Icosahedral Atmospheric Model (NICAM). SOLA 5:121-124

86. Kubokawa H, Fujiwara M, Nasuno T, Satoh M (2010) Analysis of the tropical tropopause layer using the Nonhydrostatic Icosahedral

Atmospheric Model (NICAM): aqua planet experiments. J Geophys Res 115:D08102 doi:10.1029/2009JD012686

87. Kubokawa H, Fujiwara M, Nasuno T, Miura H, Yamamoto M, Satoh M (2012) Analysis of the tropical tropopause layer using the Nonhydrostatic Icosahedral Atmospheric Model (NICAM): 2. An experiment under the atmospheric conditions of December 2006 to January 2007. J Geophys Res 117:D17114

88. Lamarque JF, Shindell DT, Josse B, Young PJ, Cionni I, Eyring V, Bergmann D, Cameron-Smith P, Collins WJ, Doherty R, Dalsoren S, Faluvegi G, Folberth G, Ghan SJ, Horowitz LW, Lee YH, MacKenzie IA, Nagashima T, Naik V, Plummer D, Righi M, Rumbold ST, Schulz M, Skeie RB, Stevenson DS, Strode S, Sudo K, Szopa S, Voulgarakis A, Zeng G (2013) The Atmospheric Chemistry and Climate Model Intercomparison Project (ACCMIP): overview and description of models, simulations and climate diagnostics. Geosci Model Dev 6:179-206

89. Laprise R (2008) Regional climate modelling. J Comp Phys 227:641-3666

90. Law RM, Peters W, Rödenbeck C, Aulagnier C, Baker I, Bergmann DJ, Bousquet P, Brandt J, Bruhwiler L, Cameron-Smith PJ, Christensen JH, Delage F, Denning AS, Fan S, Geels C, Houweling S, Imasu R, Karstens U, Kawa SR, Kleist J, Krol MC, Lin SJ, Lokupitiya R, Maki T, Maksyutov S, Niwa Y, Onishi R, Parazoo N, Patra PK, Pieterse G, et al. (2008) TransCom model simulations of hourly atmospheric CO_2: experimental overview and diurnal cycle results for 2002. Global Biogeochem Cycles 22:GB3009 doi:10.1029/2007GB003050

91. Lien G-Y, Kalnay E, Miyoshi T (2013) Effective assimilation of global precipitation: simulation experiments. Tellus A 65:19915 [http://dx.doi.org/10.3402/tellusa.v65i0.19915] *webcite* http://dx.doi.org/10.3402/tellusa.v65i0.19915

92. Lim K-SS, Hong S-Y (2010) Development of an effective double-moment cloud microphysics scheme with prognostic cloud condensation nuclei (CCN) for weather and climate models. Mon Wea Rev 138:1587-1612

93. Lin SJ (2004) A "vertically Lagrangian" finite-volume dynamical core for global models. Mon Wea Rev 132:2293-2307

94. Lin Y-L, Farley RD, Orville HD (1983) Bulk parameterization of the snow field in a cloud model. J Climate Appl Meteor 22:1065-1092

95. Liu P, Satoh M, Wang B, Fudeyasu H, Nasuno T, Li T, Miura H, Taniguchi H, Masunaga H, Fu X, Annamalai H (2009) A MJO simulated by the NICAM at 14-km and 7-km resolutions. Mon Wea Rev 137:3254-3268

96. Lohmann U, Feichter J (2005) Global indirect aerosol effects: a review. Atmos Chem Phys 5:715-737

97. Machida T, Matsueda H, Sawa Y, Nakagawa Y, Hirotani K, Kondo N, Goto K, Nakazawa T, Ishikawa K, Ogawa T (2008) Worldwide measurements of atmospheric CO_2 and other trace gas species using commercial airlines. J Atmos Oceanic Technol 25:1744-1754

98. Madden RA, Julian PR (1971) Description of a 40–50 day oscillation in the tropics. J Atmos Sci 28:702-708

99. Madden RA, Julian PR (1972) Description of global-scale circulation cells in the tropics with a 40–50 day period. J Atmos Sci 29:1109-1123

100. Majewski D, Liermann D, Prohl P, Ritter B, Buchhold M, Hanisch T, Paul G, Wergen W (2002) The operational global icosahedral-hexagonal gridpoint model GME: description and high-resolution tests. Mon Wea Rev 130:319-338

101. Mapes B, Tulich S, Nasuno T, Satoh M (2008) Predictability aspects of global aqua-planet simulations with explicit convection. J Meteor Soc Japan 86A:175-185

102. Masuda Y, Ohnishi H (1986) An integration scheme of the primitive equation model with an icosahedral-hexagonal grid system and its application to the shallow water equations. In: Short- and Medium-Range Numerical Weather Prediction, Collection of Papers Presented at the WMO/IUGG NWP Symposium. Japan Meteorological Society, Tokyo. pp 317-326

103. Masunaga H, Satoh M, Miura H (2008) A joint satellite and global cloud-resolving model analysis of a Madden-Julian Oscillation event: model diagnosis. J Geophys Res 113:D17210 doi:10.1029/2008JD009986

104. Matsueda H, Machida T, Sawa Y, Nakagawa Y, Hirotani K, Ikeda H, Kondo N, Goto K (2008) Evaluation of atmospheric CO_2 measurements from new flask air sampling of JAL airliner observations. Pap Meteorol Geophys 59:1-17

105. McGregor JL (1996) Semi-Lagrangian advection on conformal-cubic grids. Mon Wea Rev 124:1311-1322

106. Mellor GL, Yamada T (1982) Development of a turbulence closure model for geophysical fluid problems. Rev Geophys Space Phys 20:851-875

107. Meyers MP, Walko RL, Harrington JY, Cotton WR (1997) New RAMS cloud microphysics parameterization. Part II: the two-moment scheme. Atmos Res 45:3-39

108. Milbrandt JA, McTaggart-Cowan M (2010) Sedimentation-induced errors in bulk microphysics schemes. J Atmos Sci 67:3931-3948

109. Milbrandt JA, Yau MK (2005) A multimoment bulk microphysics parameterization. Part I: analysis of the role of the spectral shape parameter. J Atmos Sci 62:3051-3064

110. Milbrandt JA, Yau MK (2005) A multimoment bulk microphysics parameterization. Part II: a proposed three-moment closure and scheme description. J Atmos Sci 62:3065-3081

111. Milbrandt JA, Yau MK (2006) A multimoment bulk microphysics parameterization. Part IV: sensitivity experiments. J Atmos Sci 63:3137-3159

112. Mittal R, Iaccarino G (2003) Immersed boundary methods. Annu Rev Fluid Mech 37:239-261

113. Miura H (2007) An upwind-biased conservative advection scheme for spherical hexagonal-pentagonal grids. Mon Wea Rev 135:4038-4044

114. Miura H (2013) An upwind-biased conservative transport scheme for multi-stage temporal integrations on spherical icosahedral grids. Mon Wea Rev 141:4049-4068

115. Miura H, Skamarock WC (2013) An upwind-biased transport scheme using a quadratic reconstruction on spherical icosahedral grids. Mon Wea Rev 141:832-847

116. Miura H, Tomita H, Nasuno T, Iga S, Satoh M, Matsuno T (2005) A climate sensitivity test using a global cloud resolving model under an aqua planet condition. Geophys Res Lett 32:L19717 doi:1029/2005GL023672

117. Miura H, Satoh M, Tomita H, Nasuno T, Iga S, Noda AT (2007) A short-duration global cloud-resolving simulation with a realistic land and sea distribution. Geophys Res Lett 34:L02804 doi:10.1029/2006GL027448

118. Miura H, Satoh M, Nasuno T, Noda AT, Oouchi K (2007) A Madden-Julian Oscillation event realistically simulated by a global cloud-resolving model. Science 318:1763-1765

119. Miura H, Satoh M, Katsumata M (2009) Spontaneous onset of a Madden-Julian oscillation event in a cloud-system-resolving simulation. Geophys Res Lett 36:L13802 doi:10.1029/2009GL039056

120. Miyakawa T, Takayabu YN, Nasuno T, Miura H, Satoh M, Moncrieff MW (2012) Convective momentum transport by rainbands within a Madden-Julian oscillation in a global nonhydrostatic model with explicit deep convective processes. Part I: methodology and general results. J Atmos Sci 69:1317-1338

121. Miyakawa T, Satoh M, Miura H, Tomita H, Yashiro H, Noda AT, Yamada Y, Kodama C, Kimoto M, Yoneyama K (2014) Madden-Julian oscillation prediction skill of a new-generation global model demonstrated using a supercomputer. Nat Commun 5:3769 doi:10.1038/ncomms4769

122. Miyamoto Y, Kajikawa Y, Yoshida R, Yamaura T, Yashiro H, Tomita H (2013) Deep moist atmospheric convection in a sub-kilometer global simulation. Geophys Res Lett 40:4922-4926

123. Miyoshi T, Aranami K (2006) Applying a four-dimensional local ensemble transform Kalman filter (4D-LETKF) to the JMA Nonhydrostatic Model (NHM). SOLA 2:128-131

124. Miyoshi T, Kunii M (2011) The local ensemble transform Kalman filter with the weather research and forecasting model: experiments with real observations. Pure Appl Geophys 169:321-333

125. Miyoshi T, Yamane S (2007) Local ensemble transform Kalman filtering with an AGCM at a T159/L48 resolution. Mon Wea Rev 135:3841-3861

126. Mizuta R, Oouchi K, Yoshimura H, Noda A, Katayama K, Yukimoto S, Hosaka M, Kusunoki S, Kawai H, Nakagawa M (2005) 20-km-mesh global climate simulations using JMA-GSM model - mean climate states. J Meteor Soc Japan 84:165-185

127. Moeng CH, LeMone MA, Khairoutdinov M, Krueger SK, Bogenschutz PA, Randall DA (2009) The tropical marine boundary layer under a deep convection system: a large-eddy simulation study. J Model Earth Systems 1:16 doi:10.3894/JAMES.2009.1.16

128. Morrison H, Gettelman A (2008) A new two-moment bulk stratiform cloud microphysics scheme in the Community Atmospheric Model, version 3 (CAM3). Part I: description and numerical tests. J Climate 21:3642-3659

129. Morrison H, Curry JA, Khvorostyanov VI (2005) A new double-moment microphysics parameterization for application in cloud and climate models. Part I: description. J Atmos Sci 62:1665-1677

130. Muroi C, Toyoda E, Yoshimura H, Hosaka M, Sugi M (2002) Standard coding rule. Tenki 49:91-95 (in Japanese)

131. Nagashima T, Ohara T, Sudo K, Akimoto H (2010) The relative importance of various source regions on East Asian surface ozone. Atmos Chem Phys 10:11305-11322

132. Nakajima T, Tsukamoto M, Tsushima Y, Numaguti A, Kimura T (2000) Modeling of the radiative process in an atmospheric general circulation model. Appl Opt 39:4869-4878

133. Nakajima TY, Suzuki K, Stephens GL (2010) Droplet growth in warm water clouds observed by the A-Tran Part 1: sensitivity analysis of the MODIS-derived cloud droplet sizes. J Atmos Sci 67:1884-1896

134. Nakanishi M, Niino H (2006) An improved Mellor-Yamada level-3 model: its numerical stability and application to a regional prediction of advection fog. Boundary-Layer Meteor 119:397-407

135. Nakanishi M, Niino H (2009) Development of an improved turbulence closure model for the atmospheric boundary layer. J Meteor Soc Japan 87:895-912

136. Nakazawa T (1988) Tropical super clusters within intraseasonal variations over the western Pacific. J Meteor Soc Japan 66:823-839

137. Nasuno T (2008) Equatorial mean zonal wind in a global nonhydrostatic aquaplanet experiment. J Meteor Soc Japan 86A:219-236

138. Nasuno T (2013) Forecast skill of Madden-Julian Oscillation events in a global nonhydrostatic model during the CINDY2011/DYNAMO observation period. SOLA 9:69-73

139. Nasuno T, Satoh M (2011) Properties of precipitation and in-cloud vertical motion in a global nonhydrostatic aquaplanet experiment. J Meteor Soc Japan 89:413-439

140. Nasuno T, Satoh M (2011) Statistical relationship between maximum vertical velocity and surface precipitation of Tropical convective clouds in global nonhydrostatic aquaplanet experiment. J Meteor Soc Japan 89:553-561

141. Nasuno T, Tomita H, Iga S, Miura H, Satoh M (2007) Multi-scale organization of convection simulated with explicit cloud processes on an aqua planet. J Atmos Sci 64:1902-1921

142. Nasuno T, Tomita H, Iga S, Miura H, Satoh M (2008) Convectively coupled equatorial waves simulated by a global nonhydrostatic experiment on an aqua planet. J Atmos Sci 65:1246-1265

143. Nasuno T, Miura H, Satoh M, Noda AT, Oouchi K (2009) Multi-scale organization of convection in a global numerical simulation of the December 2006 MJO event using explicit moist processes. J Meteor Soc Japan 87:335-345

144. Neale RB, Hoskins BJ (2001) A standard test for AGCMs including their physical parameterizations: I: the proposal. Atmos Sci Lett 1:153-155 doi:10.1006/asle.2000.0019

145. Ni kovic S, Gavrilov MB, Tosic IA (2002) Geostrophic adjustment on hexagonal grids. Mon Wea Rev 130:668-683

146. Niwa Y (2010) Numerical Study on Atmospheric Transport and Surface Source/Sink of Carbon Dioxide, Dissertation. The University of Tokyo, Tokyo.

147. Niwa Y, Tomita H, Satoh M, Imasu R (2011) A three-dimensional icosahedral grid advection scheme preserving monotonicity and consistency with continuity for atmospheric tracer transport. J Meteorol Soc Jpn 89:255-268

148. Niwa Y, Patra PK, Sawa Y, Machida T, Matsueda H, Belikov D, Maki T, Ikegami M, Imasu R, Maksyutov S, Oda T, Satoh M, Takigawa M (2011) Three-dimensional variations of atmospheric CO_2: aircraft measurements and multi-transport model simulations. Atmos Chem Phys 11:3359-13375

149. Niwa Y, Machida T, Sawa Y, Matsueda H, Schuck TJ, Brenninkmeijer CAM, Imasu R, Satoh M (2012) Imposing strong constraints on tropical terrestrial CO_2 fluxes using passenger aircraft based measurements. J Geophys Res 117:D11303 doi:10.1029/2012JD017474

150. Noda AT, Oouchi K, Satoh M, Tomita H, Iga S, Tsushima Y (2010) Importance of the subgrid-scale turbulent moist process: Cloud distribution in global cloud-resolving simulatioins. Atmos Res 96:208-217

151. Noda AT, Oouchi K, Satoh M, Tomita H (2012) Quantitative assessment of diurnal variation of tropical convection simulated by a global nonhydrostatic model without cumulus parameterization. J Climate 25:5119-5134

152. Ohfuchi W, Nakamura H, Yoshioka MK, Enomoto T, Takaya K, Peng X, Yamane S, Nishimura T, Kurihara Y, Ninomiya K (2004) 10-km mesh meso-scale resolving simulations of the global atmosphere on the Earth Simulator: Preliminary outcomes of AFES (AGCM for the Earth Simulator). J Earth Simulator 1:8-34

153. Ohno T, Satoh M (2014) On the warm core of the tropical cyclone formed near the tropopause. J Atmos Sci in press

154. Onogi K, Tsutsui H, Koide H, Sakamoto M, Kobayashi S, Hatsushika H, Matsumoto T, Yamazaki N, Kamahori H, Takahashi K, Kadokura S, Wada K, Kato K, Oyama R, Ose T, Mannoji N, Taira T (2007) The JRA-25 reanalysis. J Meteorol Soc Jpn 85:369-432

155. Oouchi K, Noda AT, Satoh M, Miura H, Tomita H, Nasuno T, Iga S (2009) A simulated preconditioning of typhoon genesis controlled by a boreal summer Madden-Julian Oscillation event in a global cloud-system-resolving model. SOLA 5:65-68

156. Oouchi K, Noda AT, Satoh M, Wang B, Xie S-P, Takahashi HG, Yasunari T (2009) Asian summer monsoon simulated by a global cloud-system-resolving model: Diurnal to intra-seasonal variability. Geophys Res Lett 36:L11815 doi:10.1029/2009GL038271

157. Oouchi K, Taniguchi H, Nasuno T, Satoh M, Tomita H, Yamada Y, Ikeda M, Shirooka R, Yamada H, Yoneyama K (2012) A Prototype Quasi Real-Time Intra-Seasonal Forecasting of Tropical Convection Over the Warm Pool Region: A new Challenge of Global Cloud-System-Resolving Model for a Field Campaign. In:

Oouchi K, Fudeyasu H (eds) Cyclones: Formation, Triggers and Control. Nova Science Publishers Inc, pp 233–248

158. Oouchi K, Satoh M, Yamada Y, Tomita H, Sugi M (2014) A hypothesis and a case-study projection of an influence of MJO modulation on boreal-summer tropical cyclogenesis in a warmer climate with a global non-hydrostatic model: a transition toward the central Pacific? Front Earth Sci 2:1doi:10.3389/feart.2014.00001 (accepted)

159. Ott E, Hunt BR, Szunyogh I, Zimin AV, Kostelich EJ, Corazza M, Kalnay E, Patil DJ, Yorke JA (2004) A local ensemble Kalman filter for atmospheric data assimilation. Tellus 56A:415-428

160. Patra PK, Law RM, Peters W, Rödenbeck C, Takigawa M, Aulagnier C, Baker I, Bergmann DJ, Bousquet P, Brandt J, Bruhwiler L, Cameron-Smith PJ, Christensen JH, Delage F, Denning AS, Fan S, Geels C, Houweling S, Imasu R, Karstens U, Kawa SR, Kleist J, Krol MC, Lin SJ, Lokupitiya R, Maki T, Maksyutov S, Niwa Y, Onishi R, Parazoo N, et al. (2008) TransCom model simulations of hourly atmospheric CO_2: analysis of synoptic scale variations for the period 2002–2003. Global Biogeochem Cycles 22:GB4013 doi:10.1029/2007GB003081

161. Patra PK, Canadell JG, Houghton RA, Piao SL, Oh N-H, Ciais P, Manjunath KR, Chhabra A, Wang T, Bhattacharya T, Bousquet P, Hartman J, Ito A, Mayorga E, Niwa Y, Raymond P, Sarma VVSS, Lasco R (2013) The carbon budget of South Asia. Biogeoscience 10:513-527

162. Phillips VTJ, Donner LJ, Garner ST (2007) Nucleation processes in deep convection simulated by a cloud-system-resolving model with double-moment bulk cloud microphysics. J Atmos Sci 64:738-761

163. Pruppacher HR, Klett JD (1997) Microphysics of Clouds and Precipitation. Kluwer Academic Publisher, Heidelberg.

164. Putman WM, Suarez M (2011) Cloud-system resolving simulations with the NASA Goddard Earth Observing System global atmospheric model (GEOS-5). Geophys Res Lett 38:L16809 doi:10.1029/2011GL048438

165. Qian J-H, Semazzi FHM, Scroggs JS (1998) A global nonhydrostatic semi-Lagrangian atmospheric model with orography. Mon Wea Rev 126:747-771

166. Randall DA (1994) Geostrophic adjustment and the finite-difference shallow-water equations. Mon Wea Rev 122:1371-1377

167. Randall DA, Heikes R, Ringler T (2000) Global Atmospheric Modeling Using a Geodesic Grid With an Isentropic Vertical Coordinate. In: General Circulation Model Development, Chapter 17. Academic Press, California London. pp 509-538

168. Randall DA, Khairoutdinov M, Arakawa A, Grabowski WW (2003) Breaking the cloud-parameterization deadlock. Bull Amer Meteor Soc 84:1547-1564

169. Redler R, Valcke S, Ritzdorf H (2009) OASIS-4 – a coupling software for next generation earth system modeling. Geoscientific Model Development Discussions 2:797-843

170. Ringler TD, Randall DA (2002) A potential enstrophy and energy conserving numerical scheme for solution of the shallow-water equations on a geodesic grid. Mon Wea Rev 130:1397-1410

171. Ringler TD, Heikes RH, Randall DA (2000) Modeling the atmospheric general circulation using a spherical geodesic grid: a new class of dynamical cores. Mon Wea Rev 128:2471-2490

172. Roh W, Satoh M (2014) Evaluation of precipitating hydrometeor parameterizations in a single-moment bulk microphysics scheme for deep convective systems over the tropical open ocean. J Atmos Sci 71:2654-2673

173. Rossow WB, Schiffe RA (1999) Advances in understanding clouds from ISCCP. Bull Am Meteorol Soc 80:2261-2287

174. Rotunno R, Emanuel KA (1987) An air-sea interaction theory for tropical cyclones. Part II: evolutionary study using a nonhydrostatic axisymmetric numerical model. J Atmos Sci 44:542-561

175. Rutledge SA, Hobbs P (1983) The mesoscale and microscale structure and organization of clouds and precipitation in midlatitude cyclones. VIII: a model for the "seeder-feeder" process in warm-frontal rainbands. J Atmos Sci 40:1185-1206

176. (2014) Annual report of Development of Seamless Chemical AssimiLation System and its Application for Atmospheric Environmental Materials (SALSA). Project of the Research Program on Climate Change Adaptation (RECCA) in Ministry of Education and Sports in Japan (MEXT). [http://157.82.240.167/~salsa/

Program/SALSA_Annual_Report_FY2013.pdf]   *webcite*
Aavailable   at   http://157.82.240.167/~salsa/Program/SALSA_
Annual_Report_FY2013.pdf. Accessed at 29 Sep 2014

177. Sato T, Miura H, Satoh M (2007) Spring diurnal cycle of clouds over Tibetan Plateau: global cloud-resolving simulations and satellite observations. Geophys Res Lett 34:L18816 doi:10.1029/2007GL030782

178. Sato T, Yoshikane T, Satoh M, Miura H, Fujinami H (2008) Resolution dependency of the diurnal cycle of convective clouds over the Tibetan Plateau in a mesoscale model. J Meteor Soc Japan 86A:17-31

179. Sato T, Miura H, Satoh M, Takayabu YN, Wang Y (2009) Diurnal cycle of precipitation in the tropics simulated in a global cloud-resolving model. J Climate 22:4809-4826

180. Satoh M (2002) Conservative scheme for the compressible non-hydrostatic models with the horizontally explicit and vertically implicit time integration scheme. Mon Wea Rev 130:1227-1245

181. Satoh M (2003) Conservative scheme for a compressible nonhydrostatic model with moist processes. Mon Wea Rev 131:1033-1050

182. Satoh M (2013) Atmospheric Circulation Dynamics and General Circulation Models. Springer-PRAXIS, Heidelberg.

183. Satoh M, Kitao Y (2013) Numerical examination of the diurnal variation of summer precipitation over southern China. SOLA 9:129-133

184. Satoh M, Matsuda Y (2009) Statistics of high-cloud areas and its sensitivity to cloud microphysics with single cloud experiments. J Atmos Sci 66:2659-2677

185. Satoh M, Tomita H, Miura H, Iga S, Nasuno T (2005) Development of a global cloud resolving model - a multi-scale structure of tropical convections. J Earth Simulator 3:11-19

186. Satoh M, Matsuno T, Tomita H, Miura H, Nasuno T, Iga S (2008) Nonhydrostatic icosahedral atmospheric model (NICAM) for global cloud resolving simulations. J Comput Phys 227:3486-3514

187. Satoh M, Inoue T, Miura H (2010) Evaluations of cloud properties of global and local cloud system resolving models using

CALIPSO and CloudSat simulators. J Geophys Res 115:D00H14 doi:10.1029/2009JD012247

188. Satoh M, Oouchi K, Nasuno T, Taniguchi H, Yamada Y, Tomita H, Kodama C, Kinter J III, Achuthavarier D, Manganello J, Cash B, Jung T, Palmer T, Wedi N (2012) The Intra-Seasonal Oscillation and its control of tropical cyclones simulated by high-resolution global atmospheric models. Clim Dyn 39:2185-2206

189. Satoh M, Iga S, Tomita H, Tsushima Y, Noda AT (2012) Response of upper clouds due to global warming tested by a global atmospheric model with explicit cloud processes. J Climate 25:2178-2191

190. Satoh M, Nihonmatsu R, Kubokawa H (2013) Environmental conditions for tropical cyclogenesis associated with African easterly waves. SOLA 9:120-124

191. Sawa Y, Machida T, Matsueda H (2012) Aircraft observation of the seasonal variation in the transport of CO_2 in the upper atmosphere. J Geophys Res 117:D05305 doi:10.1029/2011JD016933

192. Seifert A (2008) On the parameterization of evaporation of raindrops as simulated by a one-dimensional rainshaft model. J Atmos Sci 65:3608-3619

193. Seifert A, Beheng KD (2001) A double-moment parameterization for simulating autoconversion, accretion and selfcollection. Atmos Res 59–60:265-281

194. Seifert A, Beheng KD (2006) A two-moment cloud microphysics parameterization for mixed-phase clouds. Part I: model description. Meteorol Atmos Phys 92:45-66

195. Seifert A, Khain A, Pokrovsky A, Beheng KD (2006) A comparison of spectral bin and two-moment bulk mixed-phase cloud microphysics. Atmos Res 80:46-66

196. Seiki T, Nakajima T (2014) Aerosol effects of the condensation process on a convective cloud simulation. J Atmos Sci 71:833-853

197. Seiki T, Satoh M, Tomita H, Nakajima T (2014) Simultaneous evaluation of ice cloud microphysics and non-sphericity of the cloud optical properties using hydrometeor video sonde and radiometer sonde in-situ observations. J Geophys Res 119:6681-6701 doi:10.1002/2013JD021086

198. Sekiguchi M, Nakaima T (2008) A k-distribution-based radiation code and its computational optimization for an atmospheric general circulation model. J Quant Spectrosc Radiat Transfer 109:2779-2793

199. Semazzi FHM, Qian J-H, Scroggs JS (1995) A global nonhydrostatic semi-Lagrangian atmospheric model without orography. Mon Wea Rev 123:2534-2550

200. Sherwood SC, Ingram W, Tsushima Y, Satoh M, Roberts M (2010) Relative humidity changes in a warmer climate. J Geophys Res 115:D09104 doi:10.1029/2009JD012585

201. Shindell DT, Miller RL, Schmidt GA, Pandolfo L (1999) Simulation of recent northern winter climate trends by greenhouse-gas forcing. Nature 399:452-455

202. Shipway BJ, Hill AA (2012) Diagnosis of systematic differences between multiple parametrizations of warm rain microphysics using a kinematic framework. Quart J Roy Meteor Soc 138:2196-2211

203. Simmons AJ, Burridge DM (1981) An energy and angular-momentum conserving vertical finite-difference scheme and hybrid vertical-coordinates. Mon Wea Rev 109:758-766

204. Skamarock WC, Klemp JB, Duda MG, Fowler LD, Park S-H (2012) A multi-scale nonhydrostatic atmospheric model using centroid Vornoi tesselations and C-grid staggering. Mon Wea Rev 140:3090-3105

205. Sohn BJ, Nakajima T, Satoh M, Jang H-S (2010) Impact of different definitions of clear-sky flux on the determination of longwave cloud radiative forcing: NICAM simulation results. Atmos Chem Phys 10:11641-11646

206. Staniforth A, Wood N (2008) Aspects of the dynamical core of a nonhydrostatic, deep-atmosphere, unified weather and climate-prediction model. J Comput Phys 227:3445-3464

207. Steppeler J, Bitzer HW, Minotte M, Bonaventura L (2002) Nonhydrostatic atmospheric modeling using a z-coordinate representation. Mon Wea Rev 130:2143-2149

208. Steppeler J, Bitzer HW, Janjic Z, Schättler U, Prohl P, Gjertsen U, Torrisi L, Parfinievicz J, Avgoustoglou E, Damrath U (2006) Prediction of clouds and rain using a z-coordinate nonhydrostatic model. Mon Wea Rev 134:3625-3643

209. Stuhne GR, Peltier WR (1996) Vortex erosion and amalgamation in a new model of large scale flow on the sphere. J Comput Phys 128:58-81

210. Sudo K, Akimoto H (2007) Global source attribution of tropospheric ozone: long-range transport from various source regions. J Geophys Res 112:D12302 doi:10.1029/2006JD007992

211. Sudo K, Takahashi M, Kurokawa J, Akimoto H (2002) CHASER: A global chemical model of the troposphere: 1. Model description. J Geophy Res 107:4339 doi:10.1029/2001JD001113

212. Sudo K, Takahashi M, Akimoto H (2002) CHASER: a global chemical model of the troposphere 2. Model results and evaluation. J Geophys Res 107:4586

213. Suzuki K, Nakajima T, Satoh M, Tomita H, Takemura T, Nakajima TY, Stephens GL (2008) Global cloud-system-resolving simulation of aerosol effect on warm clouds. Geophys Res Lett 35:L19817 doi:10.1029/2008GL035449

214. Takata K, Emori S, Watanabe T (2003) Development of the minimal advanced treatments of surface interaction and runoff. Global and Planetary Change 38:209-222

215. Takayabu YN, Iguchi T, Kachi M, Shibata A, Kanzawa H (1999) Abrupt termination of the 1997–98 El Nino in response to a Madden-Julian oscillation. Nature 402:279-282

216. Takemura T, Okamoto H, Maruyama Y, Numaguti A, Higurashi A, Nakajima T (2000) Global three-dimensional simulation of aerosol optical thickness distribution of various origins. J Geophys Res 105:17853-17873

217. Takemura T, Nakajima T, Dubovik O, Holben BN, Kinne S (2002) Single scattering albedo and radiative forcing of various aerosol species with a global three-dimensional model. J Climate 15:333-352

218. Takemura T, Nozawa T, Emori S, Nakajima TY, Nakajima T (2005) Simulation of climate response to aerosol direct and indirect effects with aerosol transport-radiation model. J Geophys Res 110:D02202 doi:10.1029/2004JD005029

219. Takemura T, Egashira M, Matsuzawa L, Ichijo H, O'ishi R, Abe-Ouchi A (2009) A simulation of the global distribution and radiative forcing of soil dust aerosols at the Last Glacial Maximum. Atmos Chem Phys 9:3061-3073

220. Taniguchi H, Yanase W, Satoh M (2010) Ensemble simulation of cyclone Nargis by a global cloud-system-resolving model–modulation of cyclogenesis by the Madden-Julian Oscillation. J Meteor Soc Japan 88:571-591

221. Tans PP, Fung IY, Takahashi T (1990) Observational constrains on the global atmospheric CO_2 budget. Science 274:1431-1438

222. Tarantola A (2005) Inverse problem theory and methods for model parameter estimation. Soc Ind Appl Math, Philadelphia, p 342, doi:10.1137/1.9780898717921

223. Taylor M, Tribbia J, Iskandarani M (1997) The spectral element method for the shallow water equations on the sphere. J Comp Phys 130:92-108

224. Terasaki K, Tanaka HL, Satoh M (2009) Characteristics of the kinetic energy spectrum of NICAM. SOLA 5:180-183

225. Thompson G, Field PR, Rasmussen RM, Hall WD (2008) Explicit forecasts of winter precipitation using an improved bulk microphysics scheme. Part II: implementation of a new snow parameterization. Mon Wea Rev 136:5095-5115

226. Thuburn J, Ringler T, Skamarock WC, Klemp JB (2009) Numerical representation of geostrophic modes on arbitrarily structured C-grids. J Comput Phys 228:8321-8335

227. Tomita H (2008) A stretched icosahedral grid by a new grid transformation. J Meteor Soc Japan 86A:107-119

228. Tomita H (2008) New microphysical schemes with five and six categories by diagnostic generation of cloud ice. J Meteor Soc Japan 86:121-142

229. Tomita H, Satoh M (2004) A new dynamical framework of nonhydrostatic global model using the icosahedral grid. Fluid Dyn Res 34:357-400

230. Tomita H, Tsugawa M, Satoh M, Goto K (2001) Shallow water model on a modified icosahedral geodesic grid by using spring dynamics. J Comp Phys 174:579-613

231. Tomita H, Satoh M, Goto K (2002) An optimization of icosahedral grid modified by spring dynamics. J Comp Phys 183:307-331

232. Tomita H, Miura H, Iga S, Nasuno T, Satoh M (2005) A global cloud-resolving simulation: preliminary results from an aqua planet experiment. Geophys Res Lett 32:L08805 doi:10.1029/2005GL022459

233. Tomita H, Goto K, Satoh M (2008) A new approach of atmospheric general circulation model: Global cloud resolving model NICAM and its computational performance. SIAM J Sci Comp 30:2755-2776

234. Trenberth K, Fasullo FT, Kiehl J (2009) Earth's global energy budget. Bull Amer Meteor Soc 90:311-323

235. Tsuchiya C, Sato K, Nasuno T, Noda AT, Satoh M (2011) Universal frequency spectra of surface meteorological fluctuations. J Climate 24:4718-4732

236. Tsushima Y, Iga S, Tomita H, Satoh M, Noda AT, Webb M (2014) High cloud increase in a perturbed SST experiment with a global nonhydrostatic model including explicit convective processes. J Adv Model Earth Syst 06:, doi:10.1002/2013MS000301

237. Wacker U, Seifert A (2001) Evolution of rain water profiles resulting from pure sedimentation: spectral vs. parameterized description. Atmos Res 58:19-39

238. Walko RL, Cotton WR, Meyers MP, Harrington JY (1995) New RAMS cloud microphysics parameterization. Part I: the single-moment scheme. Atmos Res 38:29-62

239. Watanabe M, Emori S, Satoh M, Miura H (2009) A PDF-based hybrid prognostic cloud scheme for general circulation models. Clim Dyn 33:795-816

240. Watanabe M, Suzuki T, O'ishi R, Komuro Y, Watanabe S, Emori S, Takemura T, Chikira M, Ogura T, Sekiguchi M, Takata K, Yamazaki D, Yokohata T, Nozawa T, Hasumi H, Tatebe H, Kimoto M (2010) Improved climate simulation by MIROC 5: mean states, variability, and climate sensitivity. J Climate 23:6312-6335

241. Watanabe S, Hajima T, Sudo K, Nagashima T, Takemura T, Okajima H, Nozawa T, Kawase H, Abe M, Yokohata T, Ise T, Sato H, Kato E, Takata K, Emori S, Kawamiya M (2011) MIROC-ESM 2010: model description and basic results of CMIP5-20c3m experiments. Geosci Model Dev 4:845-872

242. Webb M, Senior C, Bony S, Morcrette JJ (2001) Combining ERBE and ISCCP data to assess clouds in the Hadley Centre, ECMWF and LMD atmospheric climate models. Clim Dyn 17:905-922

243. Wedi NP (2014) Increasing horizontal resolution in numerical weather prediction and climate simulations: illusion or panacea? Philos Trans R A 372:20130289

244. Wedi NP, Smolarkiewicz PK (2009) A framework for testing global non-hydrostatic models. Q J R Meteorol Soc 135:469-484

245. Wielicki BA, Barkstrom BR, Harrison EF, Lee RB, Smith GL, Cooper JE (1996) Clouds and the Earth's Radiant Energy System (CERES): an earth observing system experiment. Bull Amer Meteor Soc 77:853-868

246. Williamson DL, Drake JB, Hack JJ, Jakob R, Swarztrauber PN (1992) A standard test set for numerical approximations to the shallow water equations in spherical geometry. J Comp Phys 102:211-224

247. Wood N, Staniforth A, White A, Allen T, Diamantakis M, Gross M, Melvin T, Smith C, Vosper S, Zerroukat M, Thuburn J (2013) An inherently mass-conserving semi-implicit semi-Lagrangian discretization of the deep-atmosphere global non-hydrostatic equations. Q J R Meteorol Soc 140:1505-1520 doi:10.1002/qj.2235

248. Wyngaard JC (2004) Toward numerical modeling in the "Terra Incognita". J Atmos Sci 61:1816-1826

249. Xiao F, Okazaki T, Satoh M (2003) An accurate semi-Lagrangian scheme for rain drop sedimentation. Mon Wea Rev 131:974-983

250. Yamada Y, Satoh M (2013) Response of ice and liquid water paths of tropical cyclones to global warming simulated by a global nonhydrostatic model with explicit cloud microphysics. J Climate 26:9931-9945

251. Yamada Y, Oouchi K, Satoh M, Tomita H, Yanase W (2010) Projection of changes in tropical cyclone activity and cloud height due to greenhouse warming: global cloud-system-resolving approach. Geophys Res Lett 37:L07709 doi:10.1029/2010GL042518

252. Yamada Y, Oouchi K, Satoh M, Noda AT, Tomita H (2012) Sensitivity of Tropical Cyclones to Large-Scale Environment in a Global non-Hydrostatic Model With Explicit Cloud Microphysics. In: Oouchi K, Fudeyasu H (eds) Cyclones: formation, Triggers and Control. Nova Science, pp 145–159

253. Yamaura T, Kajikawa Y, Tomita H, Satoh M (2013) Possible impact of a tropical cyclone on the northward migration of the Baiu frontal zone. SOLA 9:89-93

254. Yamazaki H, Satomura T (2008) Vertically combined shaved cell method in z-coordinate for non-hydrostatic atmospheric model. Atmos Sci Lett 9:171-175

255. Yamazaki H, Satomura T (2010) Nonhydrostatic atmospheric modeling using a combined cartesian grid. Mon Wea Rev 138:3932-3945

256. Yanase W, Taniguchi H, Satoh M (2010) Environmental modulation and numerical predictability associated with the genesis of tropical cyclone Nargis (2008). J Meteor Soc Japan 88:497-519

257. Yanase W, Satoh M, Yamada H, Yasunaga K, Moteki Q (2010) Continual influences of tropical waves on the genesis and rapid intensification of typhoon Durian (2006). Geophys Res Lett 37:L08809 doi:10.1029/2010GL042516

258. Yanase W, Satoh M, Taniguchi H, Fujinami H (2012) Seasonal and intraseasonal modulation of tropical cyclogenesis environment over the Bay of Bengal during the extended summer monsoon. J Clim 25:2914-2930

259. Yanase W, Satoh M, Iga S, Chan JCL, Fudeyasu H, Wang Y, Oouchi K (2012b) Multi-Scale Dynamics of Tropical Cyclone Formations in an Equilibrium Simulation Using a Global Cloud-System Resolving Model. In: Oouchi K, Fudeyasu H (eds) Cyclones: formation, triggers and control. Nova Science, pp 221–231

260. Yasunaga K, Nasuno T, Miura H, Takayabu YN, Yoshizaki M (2013) Afternoon precipitation peak simulated in an aqua-planet global non-hydrostatic model (aqua-planet-NICAM). J Meteor Soc Japan 91A:217-229

261. Yeh K-S, Côté J, Gravel S, Méthot A, Patoine A, Roch M, Staniforth A (2002) The CMC-MRB global environmental multiscale (GEM) model. Part III: nonhydrostatic formulation. Mon Wea Rev 120:329-356

262. Yoneyama K, Katsumata M, Mizuno K, Yoshizaki M, Shirooka R, Yasunaga K, Yamada H, Sato N, Ushiyama T, Moteki Q, Seiki A, Fujita M, Ando K, Hase H, Ueki I, Horii T, Masumoto Y, Kuroda Y, Takayabu YN, Shareef A, Fujiyoshi Y, McPhaden MJ, Murty VSN, Yokoyama C, Miyakawa T (2008) MISMO field experiment in the equatorial Indian Ocean. Bull Am Meteorol Soc 89:1889-1903

263. Yoneyama K, Zhang C, Long CN (2013) Tracking pulses of the Madden-Julian Oscillation. Bull Amer Meteor Soc 94:1871-1891

264. Yoshimura H, Yukimoto S (2008) Development of a simple coupler (Scup) for earth system modeling. Pap Met Geophys 59:19-29

265. Yoshizaki M, Iga S, Satoh M (2012) Eastward propagating property of large-scale precipitation systems simulated in the coarse-resolution NICAM and an explanation of its formation. SOLA 8:21-24

266. Yoshizaki M, Yasunaga K, Iga S, Satoh M, Nasuno T, Noda AT, Tomita H (2012) Why do super clusters and Madden Julian Oscillation exist over the equatorial region? SOLA 8:33-36

267. Zängl G, Tomita H, Satoh M, Ludwig T, Linardakis L, Thuburn J, Dubos T (2011) ICOMEX: ICOsahedral-grid Models for EXascale Earth system simulations. IS-ENES Workshop, Lecce.

268. Zängl G, Reinert D, Rípodas P, Baldauf M (2014) The ICON (ICOsahedral Non-hydrostatic) modelling framework of DWD and MPI-M: Description of the non-hydrostatic dynamical core. Quart J Roy Meteor Soc, doi:10.1002/qj.2378

Cassini/VIMS Observes Rough Surfaces on Titan's Punga Mare in Specular Reflection

Jason W Barnes[1], Christophe Sotin[2], Jason M Soderblom[3], Robert H Brown[5], Alexander G Hayes[8], Mark Donelan[10], Sebastien Rodriguez[6], Stéphane Le Mouélic[7], Kevin H Baines[9], and Thomas B McCord[4]

[1]Department of Physics, University of Idaho, Moscow 83844-0903, Idaho, USA

[2]Jet Propulsion Laboratory, Caltech, Pasadena 91109, California, USA

[3]Department of Earth, Atmospheric, and Planetary Sciences, Massachusetts Institute of Technology, Cambridge 02141, MA, USA

[4]Bear Fight Institute, Winthrop 98862, Washington, USA

[5]Lunar and Planetary Laboratory, University of Arizona, Tucson 85721, Arizona, USA

[6]Laboratoire AIM, Université Paris Diderot/CEA Irfu/CNRS, Centre de l'orme des Mérisiers, bât. 709, Gif/Yvette Cedex 91191, France

[7]Laboratoire de Planétologie et Géodynamique, CNRS UMR6112, Université de Nantes, Nantes, France

[8]Department of Astronomy, Cornell University, Ithaca 14853, NY, USA

[9]Space Science and Engineering Center, University of Wisconsin, Madison 53706, WI, USA

[10]University of Miami, Miami 33149, FL, USA

ABSTRACT

Cassini/VIMS high-phase specular observations of Titan's North Pole during the T85 flyby show evidence for isolated patches of rough liquid surface within the boundaries of the sea Punga Mare. The roughness shows typical slopes of $6°\pm1°$. These rough areas could be either wet mudflats or a wavy sea. Because of their large areal extent, patchy geographic distribution, and uniform appearance at low phase, we prefer a waves interpretation. Applying theoretical wave calculations based on Titan conditions our slope determination allows us to infer winds of 0.76 ± 0.09 m/s and significant wave heights of 2^{+2}_{-1} cm at the time and locations of the observation. If correct, these would represent the first waves seen on Titan's seas, and also the first extraterrestrial sea-surface waves in general.

BACKGROUND

Saturn's moon Titan posesses the only known open surface liquids beyond Earth [1]. Those liquids take the form of lakes made primarily of methane, ethane, and dissolved nitrogen [2]. The bulk of the volume of Titan's liquids occurs near the north pole [3],[4], though isolated lakes have also been observed near the south pole at Ontario Lacus [5]–[8], possibly near the equator [9], and southern mid-latitude (Sionascaig Lacus) (Vixie G, Barnes JW, Jackson B, Wilson P: Two temperate lakes on Titan. Icarus, submitted).

The existence of extraterrestrial lacustrine environments allows for the possibility of waves. In theory, expanses of liquid acted upon by sufficient winds ought to show the formation of waves on Titan as they do on Earth. The wind velocity needed to generate such waves, along with the resultant wave frequencies, will necessarily be affected by Titan's alien gravity, atmospheric density, and liquid viscosity/surface tension (which are in turn a function of composition and temperature). Theoretical calculations [10]–[12] predict that the first waves to be incited on Titan when the winds break the threshold should occur with wavelengths between 2.8 cm and 3.2 cm. Hayes et al. [12] show that these waves should be capillary-gravity waves — ones for which surface tension and gravity both contribute to the restoring force. Initial laboratory experiments with kerosene in a wind tunnel [13] showed that waves on hydrocarbons are both larger than waves on water and form at lower wind speeds, at least under Earth gravity and atmospheric conditions.

Despite concerted efforts, however, *Cassini* has thus far not detected any waves. Brown et al. [2] showed that near-infrared spectral determinations of the reflectivity of Ontario Lacus were consistent with a "smooth" surface. Wye et al. [14] used direct reflection of *Cassini's* RADAR off Ontario Lacus to constrain the surface roughness to be less than 3 mm(!). Barnes et al. [15] used a time-resolved specular reflection across north polar Jingpo Lacus to constrain wave angles at that time to be less than 0.15°. All of these observations are consistent with Titan lakes that are as flat as a millpond at the time of the observations: entirely wave-free [12].

These nondetections notwithstanding, there is indirect evidence to support the hypothesis that waves do form on Titan's lakes and seas. Wall et al. [16] claimed geomorphological evidence for waves, suggesting that the eastern shore of Ontario Lacus represents a beach formed by wave-deposited sediments. Lorenz et al. [17] explored possible explanations for why no waves have been seen, suggesting a seasonal effect, i.e. that the winds above Titan's lakes and seas were too low at the time of the observations to initiate wave formation. Calculations suggest that the threshold wind speed for wave formation under Titan conditions might be between 0.4 m/s and 0.8 m/s [12],[17]. General Circulation Models (GCMs) generally predict that the winds near Titan's north pole should have been rather quiescent until late northern spring, consistent with the low-wind explanation for Titan's lack of waves thus far [17],[18].

In this paper, we report evidence for waves on Titan's northern sea, Punga Mare. In the 'Observation' section we describe the *Cassini*

Visual and Infrared Mapping Spectrometer (VIMS) data that we interpret to show the waves in specular reflection. We describe our model for simulating the appearance of roughness-driven specular reflections away from the specular point in the 'Model' section. In the 'Analysis' section we apply that model to the VIMS observations to derive wave properties. Then, in the 'Discussion' section, we consider the implications of the discovered waves, before concluding.

OBSERVATION

In Figure 1 we show VIMS [19] cube CM_1721848119_1, acquired during the T85 flyby on 2012 July 24. It shows a very bright specular reflection (in white in Figure 1 off a lake at 87.5°N called Kivu Lacus, which was previously used to derive an atmospheric transmission spectrum [20]). This cube was acquired from a range to Titan closer than any other specular reflection observation to date (30000 km). It is this close range that makes the reflection so bright [21].

Figure 1: Cube CM_1721848119_1.*Cassini* Visual and Infrared Mapping Spectrometer (VIMS) cube CM_1721848119_1 from the T85 Titan

flyby on 2012 July 24. This cube shows a bright specular reflection (sunglint) off Kivu Lacus. This interpolated color version uses 2.0 µm as blue, 2.8 µm as green, and 5.0 µm as red. The complex structure surrounding the central glint is described in Barnes et al. ([20]).

While the Kivu specular reflection can be seen at 2.0 µm, 2.7 µm, and 2.8 µm as well, at 5 µm it is so bright that we can see additional effects over and above the raw glint (Figure 2). The specular reflection appears extended at 5 µm because of forward-scattering of specularly-reflected sunlight from haze in the atmosphere above the lake [20]. The central specular pixel and its neighbor are saturated at 5 µm and 2.8 µm. Figure 2 also shows an unusually high (though unsaturated) signal in a few pixels that are a considerable distance away from the specular point (cyan arrows in Figure 2). We show a diagram of the geometry of the T85 Kivu observation in Figure 3.

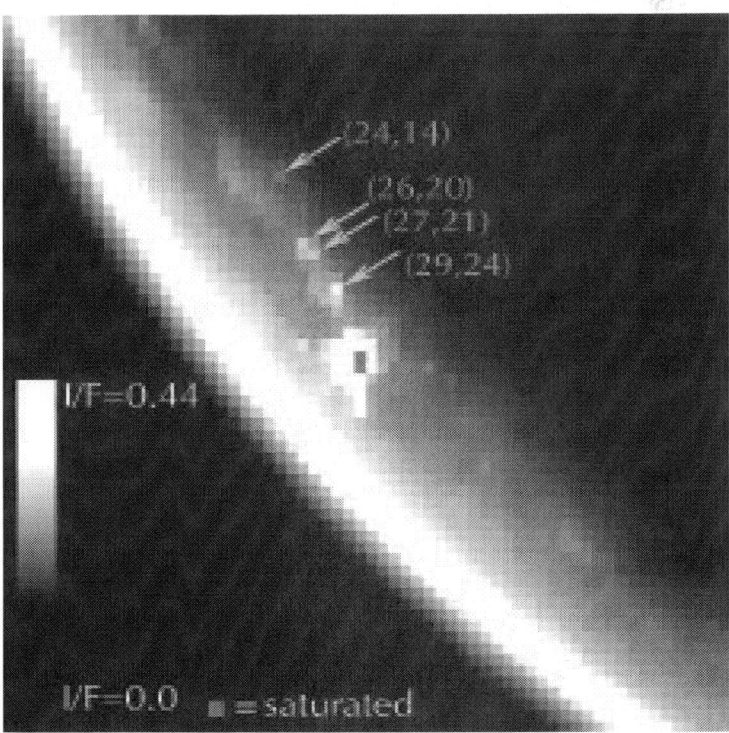

Figure 2: 5- µm image. Here we show the 5- µm window of cube CM_1721848119_1, scaled from $I/F=0.0$ to $I/F=0.44$. Red indicates

the saturation of pixels, which occurs at the primary specular reflection off Kivu Lacus. The arrows indicate the areas of interest for this paper, which show specular reflections on Punga Mare away from the specular point that may represent wave activity.

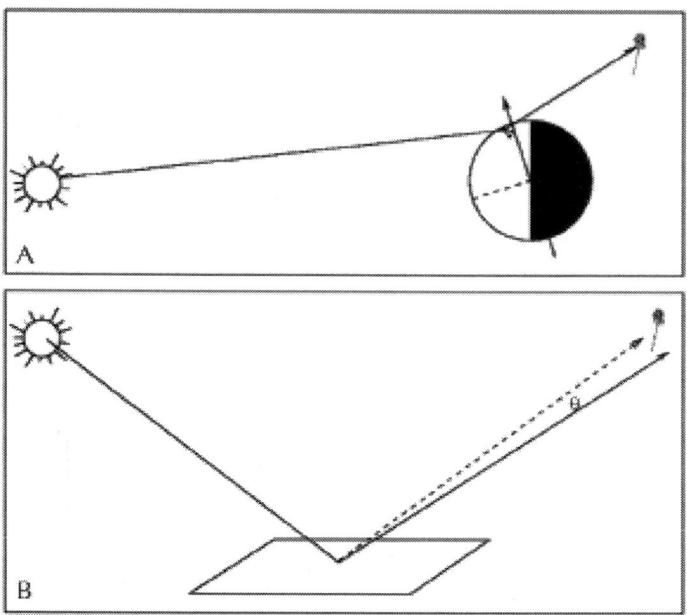

Figure 3: Observation geometry. This cartoon illustrates the geometry of the T85 VIMS observation. At top (Panel A) is the large-scale geometry with the Sun at left and *Cassini* at right. At the bottom (Panel B) is the geometry for a given pixel. Distances are not to scale in either diagram.

To place the off-specular-point areas of high signal into context, we show a mapped version of the T85 data for cube CM_1721848119_1 along with a more recent fine-resolution view from low emission angle that VIMS acquired on T94 (2013 September 12) in Figure 4. Of particular interest to note is that, in contrast to how Titan's lakes and seas appear at low phase angle, in the T85 CM_1721848119_1 high phase angle (148°) observations Titan's liquid expanses (lakes and seas) appear brighter than the surrounding terrain. This contrast inversion occurs because VIMS is seeing a specular reflection of the (somewhat) bright sky from the lake and sea surfaces [6],[17]. We show a practical demonstration of this effect in Figure 5.

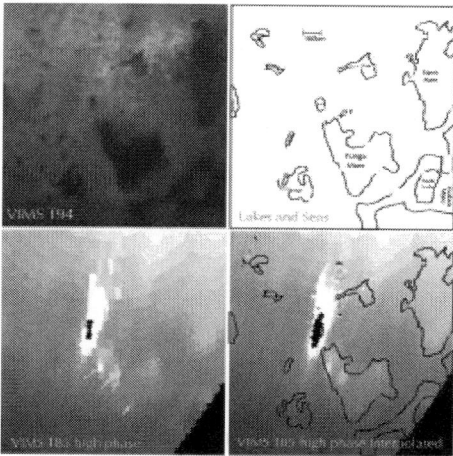

Figure 4: North polar projections. Four different orthographic projections of Titan's north polar region. At top-left is a near-infrared color version acquired on T94 with 5 µm as red, 2.0 µm as green, and 1.3 µm as blue. At upper-right we show the outlines of various named features discussed in the text. A 5 µm version of cube CM_1721848119_1 from Figure 2 is shown Lanczos interpolated at lower-right, and as pixels in the lower-left. Features with just one name are designated 'Lacus'. For purposes of these images, saturated pixels are assigned a highly negative I/F to avoid contamination when used in the interpolation.

Figure 5: Lakes in twilight. Photo looking west out the left-hand window of an aircraft on approach to Minneapolis/St. Paul airport. The Sun is just setting.

The dark areas seen on the ground are solid land. The bright areas are lakes that are specularly reflecting bright ambient sky illumination.

The bright lakes and seas do not all show the same measured I/F. There are three reasons for this. The first is that at 5 μm Titan's atmosphere is optically thin. So if you were standing on the surface in a boat on one of these lakes you would see that the sky was brighter near the horizon than at the zenith. Because lower emission angles specularly reflect a portion of the sky nearer the zenith, those areas necessarily show a dimmer specular sky reflection. The second reason is that the efficiency of the specular reflection decreases as the emission angle decreases [21]. Finally, the T85 view encompasses the terminator, meaning that past 90° incidence angle there is no direct flux at all in the lower atmosphere.

However all three of these effects are continuous and would not lead to any liquid-filled areas differing significantly in brightness from neighboring areas in a discontinuous manner. In particular, bright specular sky reflections seen in Ligeia Mare and Kraken Mare are continuous, without spurious brighter or darker areas within the seas. Kraken does have a darker area corresponding to the island Mayda Insula. Smaller lakes Sparrow Lacus, Waikare Lacus, and Muggel Lacus each are detectably brighter than their surroundings. Neagh Lacus is beyond the terminator.

The areas shown with cyan arrows in Figure 2 indicate three separate areas within Punga Mare with discontinuous and anomalously bright I/F. Two of these are at the northern end of Punga, and one is in the south. To evaluate the nature of these anomalously bright pixels, we show their spectra in Figure 6. The spectral signature shows that the bright pixels are bright only within the 5 μm window. This pattern matches the expected distribution for a specular reflection viewed through Titan's hazy atmosphere [21],[22]. VIMS does not have polarization capability, so all fluxes are total light intensity.

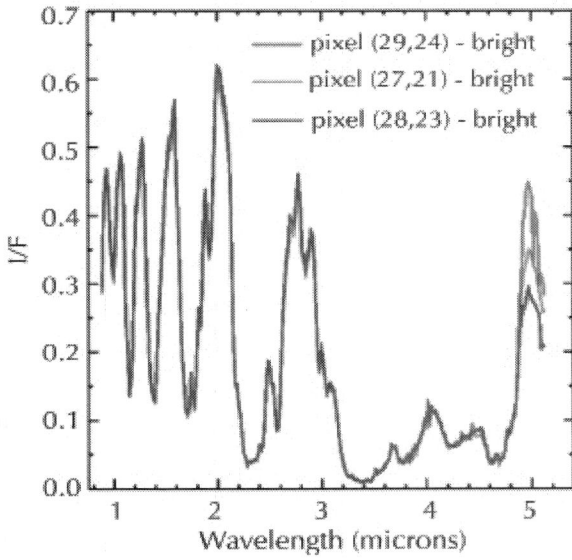

Figure 6: Near-infrared spectrum. VIMS Near-IR spectrum of three pixels from cube CM_1721848119_1 Two of the bright pixels from the cyan arrows in Figure 2 are shown in red and green, and a nearby reference pixel is in blue. The spectral signature whereby the bright areas are brighter only in the 5 μm window is consistent with their being due to specular reflection.

In particular, these spots' spectra are not consistent with an isolated patch of fog or other atmospheric aerosols above the lake — such a patch would be expected to show significant signal at wavelengths shorter than 5 μm as well [23]–[25]. While lakes and seas can reflect the image of background clouds, VIMS imaging in the wings of the 2 micron window show no evidence of cloud activity in the area at the time of the T85 observation. Hence while we cannot be certain that the spurious 5-micron flux derives from surface specular reflection, all of our tests are consistent with a surface specular phenomenon.

How can this flux be specular in origin, though, if the pixels themselves are not at the specular point on Titan's surface? Calculation of the surface specular point assumes that the surface conforms to a local equipotential surface. If the surface is instead tilted, or if portions or facets of the surface are so tilted, then either it or some of its facets can achieve a specular geometry away from the nominal specular point

We show an illustration of this effect in Figure 7. Figure 7C shows a tight specular reflection of the Sun from a smooth surface, while Figure 7B shows a broad specular reflection from a wetted rough surface. The specular reflection from the rough surface extends over several degrees, but is most intense when the surface is not just moist, but is actually covered by a thin layer of liquid.

Figure 7: A sidewalk experiment. Different types of specular reflection as viewed on a sidewalk on the University of Idaho campus. (A) Solid surfaces can produce specular reflections if their surfaces are smooth at relevant wavelength provided that they are oriented in the correct direction to specularly reflect the Sun into the observer's direction. In this case, each little arrow points to a bright speckle coming from a smooth piece of gravel sticking out of the concrete. (B) Here water has been dumped onto the surface. The water in this case follows the rough contours of the concrete, but the water's surface tension keeps it smooth at visible wavelengths and the index of refraction allows for a strong specular reflectivity. Hence there is a specular reflection, but a broad one due to some angled facets having specular geometry even away from the central specular point. (C) This image shows a specular reflection from perfectly smooth water inside the first-author's *Kepler* coffee mug. The specular signal is concentrated at a single point, and is particularly bright when compared to the weak signal from the solid surface in A.

We thus suggest that the bright, specular pixels within Punga Mare in Figure 2 could represent a rough, wet surface at those particular locations.

MODEL

We develop a numerical model of planetary specular reflections to evaluate whether the brightened areas in Figure 2 could plausibly result from specular reflection from a rough, wet surface. We use the SPICE package [26] combined with a downhill simplex numerical minimization algorithm [27] to calculate the precise orientation for a specular facet at any given point on Titan's surface given a specific observation geometry. For each latitude/longitude point, we vary both the angular deviation from zenith (θ) and azimuthal orientation (φ) for which the angular distance between the specular vector and the Sun direction vector is zero.

In Figure 8 we show the result of such calculations for θ as applied to the specific geometry in VIMS cube CM_1721848119_1. Due to the nature of specular reflection and the close observation geometry, the angular deviations required are roughly symmetric in the azimuthal direction around Titan's disk, but not in the radial direction. Furthermore, the radial and azimuthal directions are not symmetric with one another, either.

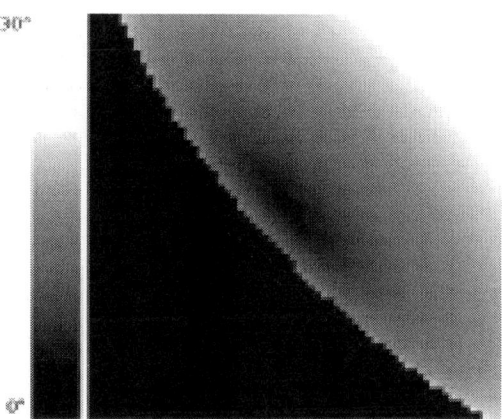

Figure 8: Surface facet angular deviation. This image, which corresponds to the pixel geometry from cube CM_1721848119_1 as shown in Figure 1, represents how far the orientation of a surface facet would have to be tilted at each location in order to achieve a specular geometry with the Sun. This value, which we call the specular facet devia-

tion θ, varies across Punga Mare from 3.4° at the end nearest to Kivu Lacus to 10° at the far end.

To then calculate the expected brightness in I/F at each point under the roughened liquid specular scenario, we calculate the fraction of randomly oriented facets that would achieve specularity with some portion of the extended solar disk. This calculation requires an assumption for the distribution of the orientation of facets within the pixel. For this work (as in [15]) we assume a two dimensional Gaussian distribution with varying widths σ. Although this assumption about the distribution may not provide the highest fidelity, given our modest 4 pixels we elect to leave a more realistic distribution including wind directionality to future work as data warrant.

Figure 9 shows the result of one such calculation for a particular pixel, assuming that width of the typical facet deviation θ distribution is $\sigma=2°$. The brightness in this image represents the relative probability for a facet to be oriented in any given direction from the zenith (θ) and azimuthally (φ) with $\theta=0.0°$ at the center of the image. The gold circles represent lines where θ is an integral multiple of σ. The white oval in the lower-left quadrant represents those facets for which the specular direction points within the Sun's disk (for the purposes of illustration the Sun's angular diameter has been multiplied by 10 for visibility).

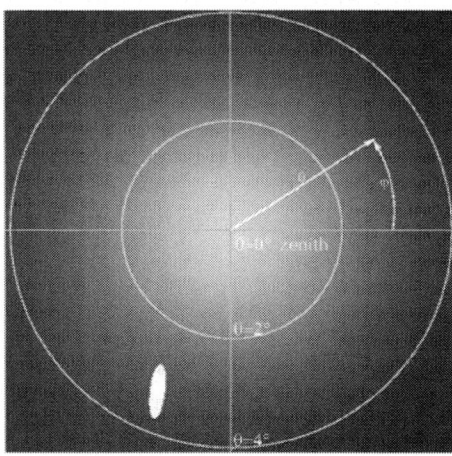

Figure 9: Modeling facet directions. Synthetic map in θ- φ space showing the relative probability of an individual surface facet facing in any

particular direction for a single particular location on Titan's surface during the T85 CM_1721848119_1 observation. This map uses $\sigma=2°$. The white oval represents the area over which the specular vector from the surface intersects the solar disk. For this particular figure the size of the solar disk has been exaggerated by a factor of 10 for visibility.

The fraction of facets f that should show specular reflection at each pixel is equal to

$$f\left(\theta,\varphi\right) = \int_0^\pi \int_{-\pi}^\pi G\left(\theta\right)\Gamma\left(\theta,\varphi\right)\ 2\pi\theta d\varphi\ \ d\theta$$

(1)

Where G is a two-dimensional gaussian function equal to

$$G\left(\theta\right) = \frac{1}{2\pi\sigma^2} e^{-\frac{\theta^2}{\sigma^2}}$$

(2)

and $\Gamma(\theta,\varphi)$ is a function equal to 1.0 when the facet at (θ,φ) corresponds to a specular direction inside the solar disk, and equal to 0.0 outside it.

For purposes of calculating the fraction of specular facets numerically, we break the solar disk into a 30-sided polygon with the vertices around the Sun's limb. We then calculate the area of that 30-sided polygon in θ-φ space and multiply the area by the Gaussian distribution value corresponding to the center of the Sun. This technique is much faster than an explicit two-dimensional numerical integral. Furthermore, given that the angular diameter of the Sun as seen from Titan ($0.05°$) is much smaller than the typical width of the facet deviation distribution that we explore ($\sigma=1-10°$) the approximation is highly accurate as well.

Finally, we normalize the total integral of the Gaussian facet deviation distribution to an assumed overall I/F value. Doing so obviates the necessity of making assumptions regarding the liquid's index of refraction and therefore of its precise composition. The final pixel value then is equal to the total solar specular flux parameter times the fraction of specular facets — this ensures that in the case where $\sigma=0°$, the specular I/F of the pixel at the specular point would be I/F_{max}. Note that because the original and model pixel I/F values are normalized, and are not true measured fluxes, summing I/F over the affected pixels does not yield I/F_{max}.

We show sample results from this model in Figure 10 for $\sigma=2°$ and $\sigma=5°$. Of particular note is the stretch in each synthetic image — for higher dispersion in specular deviation angle (higher σ) the total specular reflection becomes fuzzier, but importantly also diffuses the flux over a greater number of pixels. Hence each individual pixel shares a smaller total specular flux in the high σ case. Once we isolate the specular signal from the Punga Mare pixels, we can compare them to the model to infer surface roughness and evaluate the veracity of the roughened liquid model.

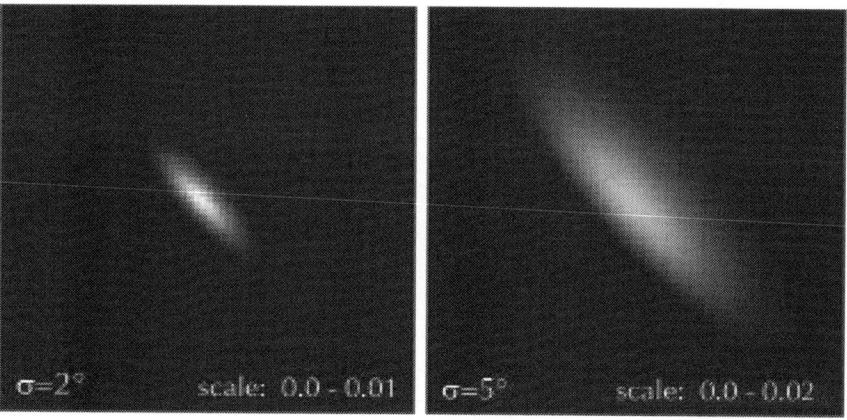

Figure 10: Model specular reflections. Synthetic images of specular flux from a uniform global ocean with Gaussian-distributed slopes of $\sigma=2°$ (left) and $\sigma=5°$ (right). These images match the viewing geometry of cube CM_1721848119_1 as shown in Figure 2.

ANALYSIS

To isolate the fraction of flux coming from the off-specular specular reflection of the Sun, we subtract an estimated background flux from each of our 4 pixels of interest. The backgrounds were taken from adjacent pixels within the lake but not showing Sun specular signal.

For pixel (29,24), closest to Kivu Lacus, we used the average of pixels (29,23) and (29,25) (see Figure 2 for location of pixels). All pixel locations are measured from the upper-left with zero-offset arrays (i.e. the first pixel is pixel number zero). For both pixel (27,21) and (26,20) we used the average of (26,19) and (27,22). And for the pixel (24,14) at the south end of Punga Mare we used the average of pixels (23,13) and (25,15). Properly subtracted and averaged over the 16 VIMS channels within Titan's 5 μm atmospheric window, we arrive at the data points shown as asterisks in Figure 11. We assume that each pixel represents a uniform patch all with the same wave facet distribution.

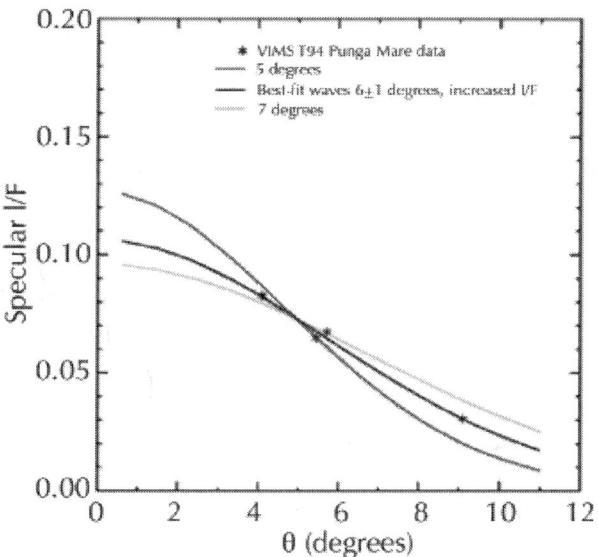

Figure 11: Wave fit. Plot of the specular flux from 4 points on Punga Mare at 5 μm from cube CM_1721848119_1 (asterisks) as a function of the specular facet deviation θ, along with a line representing the result of a χ^2 minimization fit with $\sigma=5.8°$.

To fit these data, we drove the specular brightness model with a Levenberg-Marquardt χ^2 minimization fit [27] to fit for both the maximum I/F and the facet distribution width σ. The best-fit values that result from that fit are $I/F_{max}=106\pm16$ and $\sigma=6°\pm1°$. The best-fit line (based on an assumed approximately radial profile path away from the specular point) is shown in black in Figure 11.

It is not clear why the best-fit value for I/F_{max} is discrepant from the Barnes et al. [20] value (I/F=32.4) by a factor of 3. The main specular pixel is saturated, so the error could be a result of the Barnes et al. 2013 reconstruction of the saturated values. It could also be that the true surface facet distribution does not match the 2-D Gaussian that we have assumed or that the liquid surface of Kivu Lacus where the prime specular reflection was measured is partially covered with some kind of opaque film (pond scum). Differences in the real index of refraction caused by composition differences between Kivu Lacus and Punga Mare cannot account for the offset, as at this phase angle, the first-surface reflectance of pure methane and that of pure ethane differ by only ~20%[21]. We note, however, that this new value of I/F_{max} is closer to the theoretical value as derived using the Soderblom et al. [21] relations of ~70 for methane and ~90 for ethane than that from the Kivu Lacus observation [20].

DISCUSSION

Our fit shows that the flux coming from the three isolated areas of Punga Mare studied is consistent with a rough liquid surface having characteristic slopes of 6°. Such surfaces could come about as a result of wetted mudflats, for instance, similar to the wet sidewalk in Figure 7B. A bright, dry playa surface could generate a roughened specular signal as well (see [7] Figure ten). Finally, the rough patches might also result from a purely liquid sea surface with wave activity.

The unusually high brightness of all of the areas is sufficiently high so as to only be explicable with a liquid surface: dry or even merely moist ground will not do. Dry surfaces would also not create the bright-lake effect of reflected sky brightness when seen at high emission angle as shown in Figure 5. Furthermore, the low albedo of Punga Mare could not produce a dry specular reflection of the type seen at Etosha Pan. The brightness constraint then leaves two options for the rough surfaces in Punga Mare: wet mudflats or waves.

If the liquid overlies a solid surface, then it must conform to that surface as a thin layer of liquid. This would be like the bottom part of the specular reflection shown on the sidewalk in Figure 7B. At the central specular point in Figure 7B the water has already drained downhill somewhat (toward the bottom-left in this image), leaving that part of the concrete wet but not presently covered in a layer of water. If the expanses of Punga Mare indicated by the arrows in Figure 2 represent mudflats, they must be almost entirely presently covered in sea liquid (probably a methane/ethane/nitrogen solution [2],[28]) because the specular reflection is so bright.

However, in order for that liquid surface to have the measured roughness characteristics the liquid must drape over a solid surface. Such a liquid covering over solid can occur, as evidenced by the liquid-covered sidewalk in Figure 7B. Mudflats on Earth can have very low slopes, but even then it would be difficult to achieve an appropriately thick layer of liquid over an expanse tens of kilometers long. In addition, we see the rough liquid only at discrete locations within Punga Mare. Mudflats might be expected to occur preferentially at the sea's margin, as at Ontario Lacus [6] (though in detail their distribution would depend on the sea's bathymetry). The best imaging of Punga Mare to date occurred on T94, as shown at the top left of Figure 4. Unlike the T38 Ontario Lacus data [6], the T94 data do not have the spatial resolution or signal-to-noise ratio needed to discern mudflats. Wetted floating ice [29] would be a similar solution, but would need to be similarly liquid-covered and extensive.

The other possibility is that the bright specular patches represent liquid expanses roughened by wave activity. Wind-induced waves should be possible on Titan [10]–[12],[17], even though searches until now had shown lakes and seas to be perfectly flat [14],[15]. The 6° typical slopes within the specular patches is not too dissimilar from typical slopes on Earth's oceans (4° [30]), particularly given that Hayes et al. [12] predict that Titan waves should be 7 times higher and 2 times steeper than Earth waves produced with the same wind speeds. Indeed, the angular width of sea-surface specular reflections has been used on Earth as a proxy for windspeed for 60 years[31].

We use the equations and parameters from Hayes et al. [12] and derive an explicit wavefield using the model of Donelan et al. [32], adapted to include surface tension effects (i.e., capillary-gravity waves and capillary waves). From these we calculate the expected value for both the wave angle σ and the significant wave height (defined as 4 times the root mean square (RMS) surface height) under Titan conditions (Figure 12) and assuming that the liquid viscosity is that of pure methane. Using the slope curve, we infer that our measurement of 6°±1° for surface slopes is commensurate with local winds along our line of sight and at 10 meters altitude of 0.76±0.09 m/s. This inferred windspeed is in broad agreement with the magnitude of winds expected near the threshold for the initiation of wave activity [11],[12]. From that wind determination, the expected significant wave heights for this wavefield would be 2^{+2}_{-1} cm.

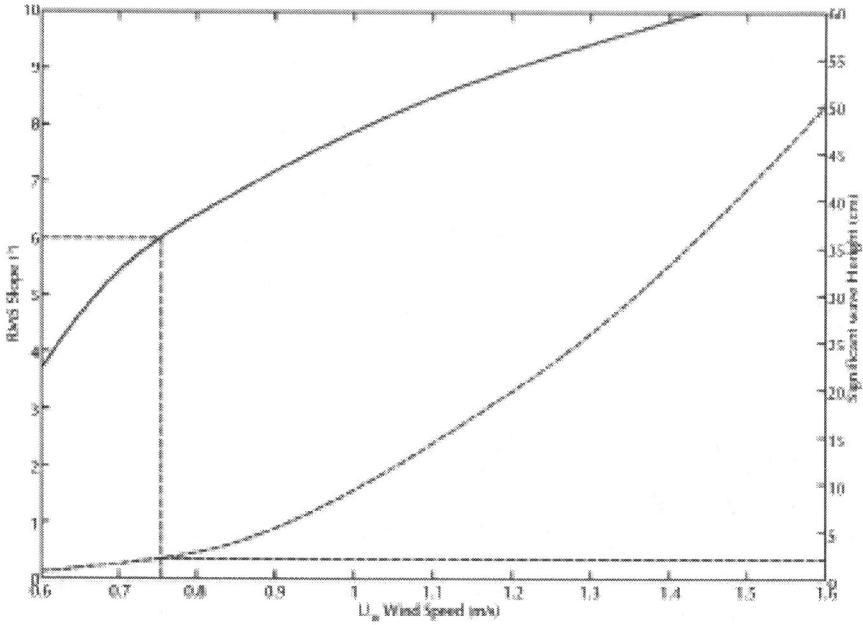

Figure 12: Slopes, winds, and wave heights. Here we plot output from the Donelan et al. [32] wave model using the Titan parameters from Hayes et al. [12]. As the solid line we plot the expected RMS surface slope (our σ) is as a function of the wind speed as measured 10 meters above the surface (U_{10}). From that relation we infer winds of 0.76±0.09 m/s in our target areas at the time of the observation. We also plot as the dashed line the significant wave height (defined to be 4 times the RMS in the liquid height across the surface) as a function of U_{10}, with the scale on the right-hand side of the plot. We therefore infer significant wave heights of 2^{+2}_{-1} cm.

Although not fully replicated by empirical studies, an experiment at 1 bar and Earth's surface gravity produced 12 mm amplitude waves with winds of 5 m/s [13]. While the wind speed associated with our

6° waves would vary with composition and viscosity, the resulting wavefield is nearly independent of composition. Hence viscosity-related systematic errors afflict the wind speed determination, the significant wave heights are mostly free from compositional systematic errors. Our calculation is in broad agreement with Lorenz et al. [33] who did a similar calculation for RADAR detectivity of waves.

The patchiness of the effect that we see represents a somewhat surprising aspect for waves. Wave activity in one portion of a lake or sea might be expected to propagate throughout the entire expanse, at least to some degree. However Hayes et al. ([12]) show that winds at or just above the threshold for wave generation could be produced locally without such propagation, since the smallest waves are quickly damped by viscous dissipation in the absence of forcing by wind. We show terrestrial analogs in Figures 13 and 14, both cases showing variable sea states (levels of wave activity) both within a single body of liquid (Great Salt Lake in Figure 13) and between adjacent lakes (Bottomless Lakes, New Mexico, USA in Figure 14).

Figure 13: great Salt Lake. Two photos taken sequentially a few seconds apart looking out the right-hand window while flying southbound on approach to

Salt Lake City airport in the photo at left, the specular point lies within an arm of the Great Salt Lake. Extending above and below the specular point is an area of enhanced wave activity, also showing specular reflection due to its surface roughness. However the area to the left of the specular reflection is wave-free. At right the specular point has moved into that wave-free zone. Now the mudflats to the upper-left of the specular point are starting to show increased specular flux due to their own roughness.

Figure 14: Bottomless lakes. Photo of specular reflections off water at Bottomless Lakes State Park, New Mexico, USA This view shows two lakes — Mirror Lake North at bottom, and Mirror Lake South at top. Mirror Lake north has smaller wave slopes, and thus exhibits a small specular reflection extending around the specular point. However the waves on Mirror Lake South in the distance are high enough so as to show significant specular flux even many degrees away from the specular point. The situation depicted may resemble the situation seen by VIMS on T85, with a relatively clean specular reflection from smooth Kivu Lacus and wave-generated roughness on Punga Mare.

Dissipation occurs more rapidly for short-wavelength capillary waves (waves with surface tension as the restoring force); it is the longer wavelength gravity waves (waves with gravity as their restoring force) that more easily propagate over long distances. Because the first waves that would result from winds just above the threshold velocity would be at the short-wavelength end of where gravity waves are possible, if those waves were incited in patches they would only propagate outside those patches with very low amplitudes.

Thus we posit that our observation may represent waves on Punga Mare that were incited by patchy winds at or just exceeding the wave-generation threshold. If correct, these would represent the first waves on open liquid detected on a body other than the Earth.

Although Global Circulation Models (GCMs) have been used to study the prospects for waves at Ligeia Mare in detail [18], no specific study has done the same for Punga Mare. In general, however, predictions show that in Titan's arctic the winds should begin to pick up as northern summer approaches [12]. We leave an investigation of whether, and under what conditions, GCMs can replicate this scenario on Punga Mare at the time in Titan's season of T85 to future work.

If real, the VIMS T85 Punga waves solve the prior paradox of waves' absence in Titan lakes and seas [17]. Indeed Titan's maria are liquid and do not have the viscosity of molasses (good for potential lake lander missions [33]). Instead, as suggested by Lorenz et al. [17], wind conditions may not have been favorable for the production of waves until recently.

We note that the patchiness of the putative waves that we see means that it would be possible for observations at a single point or even along a single chord (like [14] and [15]) to miss them. Previously unexplained variations in specular brightness on Kraken Mare on T59 [15] could be due to differing degrees of roughness across the face of the sea.

Had *Cassini* RADAR been observing Punga Mare on T85, could it have seen the putative waves that we describe here? Using the usual Synthetic Aperture RADAR (SAR) mode with high incidence angles (~30°) it would not detect these waves — in that Bragg regime their signal might be -30 or -40 db [12], well below the single-pixel noise floor of ~−20 db. At lower incidence angles, however, around 10° or less, the quasi-specular signal should make 6° waves evident even in

SAR imaging. *Cassini*'s RADAR has only observed Titan at such low incidence angles once, and in that one observation possible wave activity was observed in one location on Ligeia Mare [34]. Processing *Cassini*'s RADAR data in real aperture mode can beat down the noise and potentially reveal wave signals, but such signals would be convolved with the return from the sea floor [35].

CONCLUSIONS

VIMS T85 observations of Titan's North Pole show specular flux coming from areas within Punga Mare away from the specular point in Kivu Lacus. We develop a numerical model to simulate the appearance of a broad specular reflection off a rough surface on a spherical planet (previous work by [15] assumed very small roughness dispersion and is therefore not appropriate for moderately rough surfaces observed globally). The spectra, locations, and intensities are consistent with a surface covered in liquid and rough at wavelengths much longer than 5 μm with a typical angle of $6°\pm1°$. The inferred surface wind speeds of ~0.7 m/s are consistent with GCM predictions of increasing wind activity as northern summer approaches (e.g., [36]).

The rough patches could represent either wet mudflats or the development of waves on the sea surface. Because such mudflats would need a thin layer of liquid draped over rough mud to be consistent over areas tens of kilometers across, we prefer the waves interpretation. Future observations could definitively differentiate between the two ideas: if the regions are consistently rough as a function of time, then they are likely mudflats, whereas if the rough areas are different on a future flyby then that observation would be more consistent with waves.

The patchy nature of the putatively wavy seas implies locally variable winds near the threshold for wave generation. While future specular observations of the quality seen on T85 will be rare, there will be a few. In particular, a specular observation is planned for T101 on 2014 May 17. Combining the T85 observation with those future measurements should allow us to piece together a time-resolved picture of the frequency and intensity of high-wind events across Titan's north polar seas. The *Cassini* RADAR instrument also has prospects for detection of wave activity in the future through low-incidence synthetic aperture radar observations, bistatic experiments (T101, T102, and T106) and an altimetry pass over Kraken Mare (T104). Observations from *Cassini* or other imaging missions such as JET (Journey to Enceladus and Titan) [37] or an airplane [38] or balloon could monitor wave activity in the future by planning observations of the specular point at high phase at close enough range for the roughness effect to be seen.

AUTHORS' CONTRIBUTIONS

RHB, KHB, TBM, and CS designed and built the VIMS instrument. CS and RHB designed the observation sequence. JWB, CS, RHB, SR, and SLM analyzed the data. AGH and MD wrote and ran the Titan waves model. JWB wrote the text. All authors read and approved the final manuscript.

ACKNOWLEDGEMENTS

The authors acknowledge support from the NASA/ESA *Cassini* Project. JWB acknowledges support from NASA Cassini Data Analysis and Participating Scientists (CDAPS) grant NNX12AC28G. AGH acknowledges CDAPS grant NNX13AG03G. JMS acknowledges CDAPS grant NNX12AC25G.

REFERENCES

1. Stofan ER, Elachi C, Lunine JI, Lorenz RD, Stiles B, Mitchell KL, Ostro S, Soderblom L, Wood C, Zebker H, Wall S, Janssen M, Kirk R, Lopes R, Paganelli F, Radebaugh J, Wye L, Anderson Y, Allison M, Boehmer R, Callahan P, Encrenaz P, Flamini E, Francescetti G, Gim Y, Hamilton G, Hensley S, Johnson WTK, Kelleher K, Muhleman D, et al.: The lakes of Titan. Nature2007, 445:61–64.

2. Brown RH, Soderblom LA, Soderblom JM, Clark RN, Jaumann R, Barnes JW, Sotin C, Buratti B, Baines KH, Nicholson PD: The identification of liquid ethane in Titan's Ontario Lacus. Nature2008, 454:607–610.

3. Hayes A, Aharonson O, Callahan P, Elachi C, Gim Y, Kirk R, Lewis K, Lopes R, Lorenz R, Lunine J, Mitchell K, Mitri G, Stofan E, Wall S: Hydrocarbon lakes on Titan: distribution and interaction with a porous regolith. Geophys Res Lett2008, 35:L9204.

4. Sotin C, Lawrence KJ, Reinhardt B, Barnes JW, Brown RH, Hayes AG, Le Mouélic S, Rodriguez S, Soderblom JM, Soderblom LA, Baines KH, Buratti BJ, Clark RN, Jaumann R, Nicholson PD, Stephan K: Observations of Titan's northern lakes at 5μm: implications for the organic cycle and geology. Icarus2012, 221:768–786.

5. Turtle EP, Perry JE, McEwen AS, Del Genio AD, Barbara J, West RA, Dawson DD, Porco CC: Cassini imaging of Titan's high-latitude lakes, clouds, and south-polar surface changes. Geophys Res Lett2009, 36:L2204.

6. Barnes JW, Brown RH, Soderblom JM, Soderblom LA, Jaumann R, Jackson B, Le Mouélic S, Sotin C, Buratti BJ, Pitman KM, Baines KH, Clark RN, Nicholson PD, Turtle EP, Perry J: Shoreline features of Titan's Ontario Lacus from Cassini/VIMS observations. Icarus2009, 201:217–225.

7. Cornet T, Bourgeois O, Le Mouélic S, Rodriguez S, Lopez Gonzalez T, Sotin C, Tobie G, Fleurant C, Barnes JW, Brown RH, Baines KH, Buratti BJ, Clark RN, Nicholson PD: Geomorphological significance of Ontario Lacus on Titan: integrated interpretation of Cassini VIMS, ISS and RADAR data and comparison with the Etosha Pan (Namibia). Icarus2012, 218:788–806.

8. Cornet T, Bourgeois O, Le Mouelic S, Rodriguez S, Sotin C, Barnes JW, Brown RH, Baines KH, Buratti BJ, Clark RN, Nicholson PD: Edge detection applied to Cassini images reveals no measurable displacement of Ontario Lacus' margin between 2005 and 2010. J Geophys Res (Planets)2012, 117:E07005.

9. Griffith CA, Lora JM, Turner J, Penteado PF, Brown RH, Tomasko MG, Doose L, See C: Possible tropical lakes on Titan from observations of dark terrain. Nature2012, 486:237–239.

10. Ghafoor NAL, Zarnecki JC, Challenor P, Srokosz MA: Wind-driven surface waves on Titan. J Geophys Res2000, 105:12077–12092.

11. Lorenz RD, Hayes AG: The growth of wind-waves in Titan's hydrocarbon seas. Icarus2012, 219:468–475.

12.   Hayes AG, Lorenz RD, Donelan MA, Manga M, Lunine JI, Schneider T, Lamb MP, Mitchell JM, Fischer WW, Graves SD, Tolman HL, Aharonson O, Encrenaz PJ, Ventura B, Casarano D, Notarnicola C: Wind driven capillary-gravity waves on Titan's lakes: hard to detect or non-existent?Icarus2013, 225:403–412.

13.   Lorenz RD, Kraal ER, Eddlemon EE, Cheney J, Greeley R: Sea-surface wave growth under extraterrestrial atmospheres: preliminary wind tunnel experiments with application to Mars and Titan. Icarus2005, 175:556–560.

14.   Wye LC, Zebker HA, Lorenz RD: Smoothness of Titan's Ontario Lacus: constraints from Cassini RADAR specular reflection data. Geophys Res Lett2009, 36:L16201.

15.   [http:/ / www.sciencedirect.com/ science/ article/ B6WGF-5161PFV-4/ 2/ 7a95516e9191ed9db6df21c7bf5b97b4] Barnes JW, Soderblom JM, Brown RH, Soderblom LA, Stefan K, Jaumann R, Le Mouelić S, Rodriguez S, Sotin C, Buratti BJ, Baines KH, Clark RN, Nicholson PD: Wave constraints for Titan's Jingpo Lacus and Kraken Mare from VIMS specular reflection lightcurves. Icarus2011, 211:722–731.

16.   Wall S, Hayes A, Bristow C, Lorenz R, Stofan E, Lunine J, Le Gall A, Janssen M, Lopes R, Wye L, Soderblom L, Paillou P, Aharonson O, Zebker H, Farr T, Mitri G, Kirk R, Mitchell K, Notarnicola C, Casarano D, Ventura B: Active shoreline of Ontario Lacus, Titan: a morphological study of the lake and its surroundings. Geophys Res Lett2010, 37:L5202.

17.   Lorenz RD, Newman C, Lunine JI: Threshold of wave generation on Titan's lakes and seas: Effect of viscosity and implications for Cassini observations. Icarus2010, 207:932–937.

18.   Lorenz RD, Tokano T, Newman CE: Winds and tides of Ligeia Mare, with application to the drift of the proposed time TiME (Titan Mare Explorer) capsule. Planet Space Sci2012, 60:72–85.

19.   Brown RH, Baines KH, Bellucci G, Bibring JP, Buratti BJ, Capaccioni F, Cerroni P, Clark RN, Coradini A, Cruikshank DP, Drossart P, Formisano V, Jaumann R, Langevin Y, Matson DL, McCord TB, Mennella V, Miller E, Nelson RM, Nicholson PD, Sicardy B, Sotin C: The Cassini Visual And Infrared Mapping Spectrometer (Vims) investigation. Space Sci Rev2004, 115:111–168.

20. Barnes JW, Clark RN, Sotin C, Ádámkovics M, Apperé T, Rodriguez S, Soderblom JM, Brown RH, Buratti BJ, Baines KH, Le Mouelić S, Nicholson PD: A transmission spectrum of Titan's north polar atmosphere from a specular reflection of the sun. ApJ2013, 777:161.

21. Soderblom JM, Barnes JW, Soderblom LA, Brown RH, Griffith CA, Nicholson PD, Stephan K, Jaumann R, Sotin C, Baines KH, Buratti BJ, Clark RN: Modeling specular reflections from hydrocarbon lakes on Titan. Icarus2012, 220:744–751.

22. Stephan K, Jaumann R, Brown RH, Soderblom JM, Soderblom LA, Barnes JW, Sotin C, Griffith CA, Kirk RL, Baines KH, Buratti BJ, Clark RN, Lytle DM, Nelson RM, Nicholson PD: Specular reflection on Titan: liquids in Kraken Mare. Geophys Res Lett2010, 37:L7104.

23. Rodriguez S, Le Mouélic S, Rannou P, Tobie G, Baines KH, Barnes JW, Griffith CA, Hirtzig M, Pitman KM, Sotin C, Brown RH, Buratti BJ, Clark RN, Nicholson PD: Global circulation as the main source of cloud activity on Titan. Nature2009, 459:678–682.

24. Brown ME, Roberts JE, Schaller EL: Clouds on Titan during the Cassini prime mission: a complete analysis of the VIMS data. Icarus2010, 205:571–580.

25. Rodriguez S, Le Mouélic S, Rannou P, Sotin C, Brown RH, Barnes JW, Griffith CA, Burgalat J, Baines KH, Buratti BJ, Clark RN, Nicholson PD: Titan's cloud seasonal activity from winter to spring with Cassini/VIMS. Icarus2011, 216:89–110.

26. [http://adsabs.harvard.edu/abs/2013LPI....44.1224A] Acton CH: SPICE products available to the planetary science community. In Lunar and Planetary Institute Conference Abstracts; 1999:1233–1234. .

27. Press WH, Teukolsky SA, Vetterling WT, Flannery BP: Numerical Recipes: The Art of Scientific Computing. University Press, Cambridge; 2007.

28. Mitri G, Showman AP, Lunine JI, Lorenz RD: Hydrocarbon lakes on Titan. Icarus2007, 186:385–394.

29. [http://www.sciencedirect.com/science/article/pii/S0019103512004824] Hofgartner JD, Lunine JI: Does ice float in Titan's lakes and seas?Icarus2013. (0). []

30. [http://dx.doi.org/10.1029/98JC00780] Walsh EJ, Hagan DE, Rogers DP, Weller RA, Fairall CW, Friehe CA, Burns SP, Khelif D, Vandemark DC, Swift RN, Scott JF: Observations of sea surface mean square slope under light wind during the Tropical Ocean-Global Atmosphere Coupled Ocean-Atmosphere Response Experiment. *J Geophys Res C Oceans*1998, 103(C6):12603–12612.

31. Cox C, Munk W: Measurement of the roughness of the sea surface from photographs of the sun's glitter. J Opt Soc Am1954, 44:838–850.

32. [http://dx.doi.org/10.1029/2011JC007787] Donelan MA, Curcic M, Chen SS, Magnusson AK: Modeling waves and wind stress. *J Geophys Res C Oceans*2012, 117(C11):n/a–n/a.

33. Lorenz RD, Biolluz G, Encrenaz P, Janssen MA, West RD, Muhleman DO: Cassini RADAR: prospects for Titan surface investigations using the microwave radiometer. Planet Space Sci2003, 51:353–364.

34. Hofgartner JD, Hayes AG, Lunine JI, Zebker H, Stiles BW, Sotin C, Barnes JW, Brown RH, Encrenaz P, Kirk RD, Le Gall A, Lopes RM, Lorenz RD, Malaska M, Mitchell KL, Paillou P, Radebauch J, Turtle E, Wall S, Wood C, The Cassini RADAR Team: The discovery of transient features in a Titan sea. Nat Geosci2014, 7:493–496. doi:10.1038/ngeo2190.

35. Mastrogiuseppe M, Poggiali V, Hayes A, Lorenz R, Lunine J, Picardi G, Seu R, Flamini E, Mitri G, Notarnicola C, Paillou P, Zebker H: The bathymetry of a Titan sea. Geophys Res Lett2014, 41:1432–1437.

36. Lebonnois S, Burgalat J, Rannou P, Charnay B: Titan global climate model: a new 3-dimensional version of the IPSL Titan GCM. Icarus2012, 218:707–722.

37. [http://adsabs.harvard.edu/abs/2010DPS....42.4931S] Sotin C, Altwegg K, Brown RH, Hand K, Soderblom J, JET Team: JET: a Journey to Enceladus and Titan. In *AAS/Division for Planetary Sciences Meeting Abstracts #42, Volume 42 of AAS/Division for Planetary Sciences Meeting Abstracts*; 2010:#49.31. .

38. Barnes JW, Lemke L, Foch R, McKay CP, Beyer RA, Radebaugh J, Atkinson DH, Lorenz RD, Le Mouélic S, Rodriguez S, Gundlach J, Giannini F, Bain S, Flasar FM, Hurford T, Anderson CM, Merrison

J, Ádámkovics M, Kattenhorn SA, Mitchell J, Burr DM, Colaprete A, Schaller E, Friedson AJ, Edgett KS, Coradini A, Adriani A, Sayanagi KM, Malaska MJ, Morabito D, et al.: AVIATR — Aerial Vehicle for In-situ and Airborne Titan Reconnaissance. A Titan airplane mission concept. Exp Astron2012, 33:55–127.

Chapter 5

Early Life on Land and the First Terrestrial Ecosystems

Hugo Beraldi-Campesi

Institute of Geology, UNAM, Ciudad Universitaria, Mexico, DF 04510, Mexico

ABSTRACT

Terrestrial ecosystems have been largely regarded as plant-dominated land surfaces, with the earliest records appearing in the early Phanerozoic (<550 Ma). Yet the presence of biological components in pre-Phanerozoic rocks, in habitats as different as soils, peats, ponds, lakes, streams, and dune fields, implies a much earlier type of terrestrial ecosystems. Microbes were abundant by ~3,500 Ma ago and surely adapted to live in subaerial conditions in coastal and inland environments, as they do today. This implies enormous capacities for rapid adaptations to changing conditions, which is supported by a suggestive fossil record. Yet, evidence of "terrestrial" microbes is rare and indirect in comparison with fossils from shallow or deeper marine environments, and its record has been largely overlooked. Consequently, the notion that microbial communities may have formed the earliest land ecosystems has not been widely accepted nor integrated into our general knowledge. Currently, an ample record of shallow marine and lacustrine biota in ~3,500 Ma-old deposits, together with evidence of microbial colonization of coastal environments ~3,450 Ma ago and indirect geochemical evidence that suggests biological activity in >3,400 Ma-old paleosols endorses the idea that life on land perhaps occurred in parallel with aquatic life back in the Paleoarchean. The rapid adaptations seen in modern terrestrial microbes, their outstanding tolerance to extreme and fluctuating conditions, their early and rapid diversification, and their old fossil record collectively suggest that they constituted the earliest terrestrial ecosystems, at least since the Neoarchean, further succeeding on land and forming a biomass-rich cover with mature soils where plant-dominated ecosystems later evolved. Understanding how life diversified and adapted to non-aquatic conditions from the actualistic and paleontological perspective is critical to understanding the impact of life on the Earth's systems over thousands of millions of years.

INTRODUCTION

Definition of "terrestrial"

Habitable, non-aquatic environments must have existed all throughout the geologic history of Earth unless its surface was entirely under water, which seems unlikely. The definition of a terrestrial environment may not be as trivial as it sounds. "Terrestrial" is defined here as non-aquatic environments. However, even fully aquatic ecosystems, such as lakes and coastal environments, cover a wide spectrum of mixed environments where aquatic and non-aquatic landscapes develop and overlap over time. Habitats above sea level include aquatic (ice-covered and ice-free lakes, ponds and wetlands, peats, rivers and streams, geothermal fields) and non-aquatic environments (especially areas with low rain regimes) that experience drastic changes governed by tectonic activity and climatic conditions, including the rise and fall of sea level, glaciations, and rain regimes (e.g., Romans and Graham 2013). Microbes can be expected in all these environments and, in the long term, they may have strongly influenced the regional topography, sedimentation rates, sedimentary dynamics, and the reworking of previously emplaced materials. This might be difficult to interpret sometimes from the sedimentary record, especially in environments whose configuration and sedimentary dynamics can change in a relatively short time (days to decades), such as coastal areas (deltas, estuaries, lagoons, evaporitic flats, dunes, etc.; e.g., Hamblin and Christiansen 2007), going from aquatic to non-aquatic environments in a few centimeters or meters of rock strata.

Sedimentary deposits originating in fully aquatic environments (fluvial, lacustrine, shallow marine) can be further exposed to the atmosphere for long periods of time and undergo pedogenetic processes, which transform some of the original features of the deposit into a soil (e.g. Paul et al.2001and references therein). The rocks thus keep gross characteristics of the primary deposit but are overprinted with the in situ, secondary features derived from completely different environmental conditions. Besides, elucidating timescales at the outcrop scale is not always feasible and is frequently overlooked in less resolved regional studies. Ultimately, this has surely contributed to biases in the interpretation of the evolution of the biosphere. In this regard, the study of pedogenetic (e.g., development of horizons, hardpans or duricrusts, peds, and clay compositional and mechanical—e.g., slickensides—features, etc.) and microbial processes in spring and stream microbialites (e.g., travertines, tufas, sinters) and exposed sedimentary and rock surface habitats (endolithic habitats and cryptogamic covers), is of particular importance for the more comprehensive understanding of past life because these represent terrestrial habitats expected on ancient continental surfaces.

Caution and Re-interpretation of the Rock Record

Through the integrative study of rocks and the understanding of the processes that formed them, including the fossil record and our ability to date materials isotopically, we have built a concept of the evolution of the geosphere and the biosphere (see compendiums by Schopf 1983; Canup and Righter2000 ; Eriksson et al. 2004 ; Schieber et al. 2007; Van Kranendonk et al. 2007a ; Kasting 2009 ; Taylor et al. 2009 ; Knoll et al. 2012), despite important and ongoing debate on the details. One key element in this picture is the inception of life, which has been interacting with and changing, maintaining, and recirculating most of the materials existing in the atmosphere and the supracrustal section of the Earth for over ~80% of Earth's history. This scenario has been studied and interpreted over the years, aided by the technology available at the moment, not always correctly and also biased sometimes by the general consensus (e.g., Hallbauer 1975 ; Gray and Boucot 1994 ; Windley2007). Also, our appreciation of the timing of geological phenomena (soil formation, seafloor spreading, mountain building, rock erosion, lake succession, etc.) may be difficult to relate to other global changes (e.g. rapid and profuse volcanism and rapid climatic oscillations coexisting with slow seafloor spreading and continental drift) when interpreting the rock record.

Besides the record being incomplete, the speed at which biology operates compared to geology implies that hiatuses of tens of millions of years (negligible in Pracambrian timescales), represented by only a few centimeters or meters of rock strata, bear enormous opportunities for biological developments and adaptations that may have not been preserved. This conceptualization of the speed at which biotic and geological events occur simultaneously requires a careful examination and re-evaluation of Precambrian geological materials (aided by the advancement of scientific knowledge and technology around it) with a readiness to consider challenging ideas (e.g., Retallack 2013), especially when recognizing fossils or when trying to reconcile them with their paleoenvironments (Xiao and Knauth 2013).

The fossil record of microbes is largely related to aquatic environments, and while abundant morphological, chemical, and geochemical evidence of diverse, aquatic Archean life has reached wide acceptance and consensus, the existence of life on Precambrian lands is not always taken for granted. The historical perception of plants as the dominant group on the land, together with the first discoveries of macroscopic fossils only in Phanerozoic rocks and the inability to correctly interpret microbial and algal biosignatures, has perhaps infused the generalized understanding of "colonized" terrestrial ecosystems exclusively for plants (e.g., Bambach 1999). In some instances, even when the existence of Precambrian terrestrial ecosystems may be recognized, they are still treated doubtfully or not given adequate attention (Shear 1991 ; Behrensmeyer et al. 1992 ; DiMichele and Hook 1992 ; Gray and Shear 1992 ; Gray and Boucot 1994) even after previous and important discussions on the topic (e.g., Wright 1985 ; but see also Labandeira 2005).

The possible misinterpretation of terrestrial paleoenvironments and their relatively poor preservation in the sedimentary record does not necessarily mean that terrestrial life did not exist on the early continents. Today there is growing evidence indicative of non-aquatic environments colonized by microbes early in Earth's history, which is consistent with the extent of modern microbial life on analog "barren" lands (deserts, polar plains, alpine rocks, etc.) their outstanding diversity and metabolic capabilities, and by the great diversity and distribution of Precambrian microfossils, which is a reflection of the microbial ubiquity at that time.

The Setting for Early Life

The oldest materials yet found in the Solar System occur in meteorites and are ~4,570 Ma (Mega annum, million years) of age (Bouvier and Wadhwa 2010), which may serve as a starting point for the condensation of the first solids in our Solar System. By contrast, the oldest materials on Earth (zircon crystals) go back ~4,400 Ma (Wilde et al. 2001), leaving a hiatus of ~170 Ma in Earth's geological history. Regardless, it is assumed that the Moon was already formed before 4,400 Ma (Canup and Righter 2000 ; Yu and Jacobsen 2011) and that the Earth's nucleus, mantle, and lithosphere were already differentiated (Nelson 2004 ; Boyet and Carlson 2005). At least by ~4,200 Ma, but perhaps 200 Ma earlier, large water bodies were in place (Mojzsis et al. 2001 ; Nutman 2006; Cavosie et al. 2007 , but see alternative views by Deming 2002), while granitic (continental) and basaltic (oceanic) crusts were constantly growing, resurfacing, and remelting, interacting with water in non-uniform regimes that evolved drastically from the Hadean to the Neoarchean (Komiya et al.1999 ; Nutman et al. 2002 ; Myers 2004 ; Rino et al. 2004 ; Van Kranendonk 2004and references therein; Furnes et al. 2007a ; Adam et al. 2012), changing from plume-dominated to plate-dominated tectonics toward the late Paleoarchean

(Van Kranendonk et al. 2007b). It is plausible then, that by the end of the heavy bombardment (Gomes et al. 2005 ; Hartmann et al. 2000) some ~3,800 Ma ago, the primitive lands and oceans were open niches ready for the pioneering and rapidly evolving microscopic life forms, for which occasional "sterilizing" perturbations may be irrelevant given the resilience and time scales at which biology operates compared to geology.

Although life may have appeared only a few hundred million years after Earth's accretion (e.g., Lopez-Garcia et al. 2006), sedimentary rocks older than ~3,850 Ma (Nutman et al. 1996 ; Ishizuka2008 ; Nutman et al. 2010 ; O'Neil et al. 2011 ; Mloszewska et al. 2012), where biotic events are most likely to be imprinted, are uncommon on Earth. Yet, potential traces of life (e.g. biogenically precipitated carbonates) may even be present in this ancient record (Nutman et al. 2010), suggesting that life itself can be several million years older than the oldest known stromatolites and microfossils. Other putative biosignatures older than 3,500 Ma (carbonaceous spherules associated with apatite globules; see McKeegan et al. 2007; Papineau et al. 2010a, 2010b) are also controversial (see Myers2001 ; van Zuilen et al. 2002 ; Fedo and Whitehouse 2002 ; Papineau et al. 2011) and may not imply a syngenetic timing of formation with the host rock. Biosignatures of particular interest would be those associated with biogenic banded iron formations (e.g., Dauphas et al. 2004 ; Trendall and Blockley 2004 ; Kappler et al. 2005 ; Konhauser et al. 2005 ; Koehler et al. 2010 ; Mloszewska et al.2012) given their potential antiquity of ~4,300 Ma (O'Neil et al. 2009).

Microfossils, microbialites, and isotopic and molecular biomarkers indicate that prokaryotic life was abundant by 3,500–3,400 Ma ago in shallow and deep marine environments (Lowe 1980 ; Walter et al. 1980 ; Awramik et al. 1983 ; Schopf 1983 ; Walter 1983 ; Walsh, 1992 ; Walsh and Lowe 1985 ,1999 ; Rasmussen 2000 ; Westall et al. 2001 ; Furnes et al. 2004 ; Shen and Buick 2004 ; Tice and Lowe 2004 ; Allwood et al. 2006 ; Banerjee et al. 2006; Westall et al. 2006a, 2006b ; Ueno et al.2006 ; Schopf et al. 2007 and references therein; Shen et al. 2009 ; Westall 2010 ; Wacey et al.2011), which supports the notion that coastal and estuarine areas could have been very productive by that time and that photosynthesis was already operating (Awramik 1992 ; Rosing and Frei 2004 ; Tice and Lowe 2004 ; Buick 2008 ; Hoashi et al. 2009 ; Kato et al. 2009 ; Kendall et al. 2010), though perhaps not necessarily oxygenic (Kirschvink and Kopp 2008 ; Westall et al. 2011 ; Li et al.2012).

Many different settings have been proposed as likely or "optimum" for the emergence and prosperity of life, ranging from deep-sea hydrothermal vents and geothermal springs, to land surfaces and mineral-water-air interphases (Baross and Hoffman 1985; Retallack 1986a ; Holm 1992 ; Battistuzzi and Hedges 2009 ; Aller et al. 2010 ; Hazen and Sverjensky 2010 ; Mulkidjanian et al. 2012). However, one preferred environment where many of the oldest signs of life are found is shallow marine continental margins (see references in Schopf and Klein 1992). Whether this is a true fact or a consequence of the incompleteness/selectivity of the record is still to be resolved. However, in these coastal environments microbes were likely to have been periodically exposed and desiccated, as happens in most such environments today, and likely developed adaptations for long-term desiccation regimes (e.g., thick hygroscopic sheaths) and high UV radiation (e.g., living interstitially).

Some of the oldest examples of life activity, which come precisely from aquatic, shallow marine (Klein et al. 1987 ; Schopf and Klein 1992 ; Van Kranendonk et al. 2008 ; Westall 2010 ; Van Kranendonk2011 ; Hickman and Van Kranendonk 2012), shallow lacustrine (Awramik and Buchheim 2009 ; Hickman and Van Kranendonk 2012), and intertidal environments (e.g., Noffke et al. 2006 ; Noffke2010 ; Noffke et al. 2011 ; Westall et al. 2011), show signs of evaporation (e.g., Noffke et al. 2008 ; Westall et al. 2011 ; Hickman and Van Kranendonk 2012), which suggests that early microbial communities in shallow waters had to deal with periodic desiccation and UV radiation >3,400 Ma ago. This further implies adaptations to resist desiccation, salinity fluctuations, and UV radiation that could be successfully used even after prolonged desiccation. Dry conditions can be expected also for lacustrine and fluviatile environments. Desiccation allows dispersion by wind, which seems like a reasonable means for land colonization. Through this mode of dispersion, communities would tend to be at or near the surface instead of underground, even when migration to aquifers can occur. Perhaps environments with periodic subaerial exposure (especially estuarine and intertidal) were crucial scenarios for a biological transition from water to land.

Apparently not only prokaryotes were abundant in shallow Precambrian environments; the oldest eukaryotic-like fossils (acritarchs; Buick 2010) found so far (that perhaps needed oxygen for advantageous energetic and metabolic capabilities) are ~3,200 Ma old and were also present in estuarine environments (Javaux et al. 2010). This indicates that life achieved a relatively rapid global presence and had diversified enough (Kandler 1994 ; Altermann and Schopf 1995 ; Ueno et al. 2006; Blank 2009 ; David and Alm 2011) to occupy a wide variety of ecological niches by the Paleoarchean, even in places that may have been severely disturbed by asteroid impacts (see Walsh1992and references therein). Even greater biological diversity, ubiquity, abundance, and habitats are seen in the younger Proterozoic record (e.g., Schopf 1992a ; Schopf and Klein 1992), for which rocks are better preserved and more abundant than Archean ones.

The Fossil Record of Terrestrial Life

The earliest remnants of continental crust may derive from ≥3,500 Ma-old submillimeter zircons (Nutman 2006) and regional rock outcrops (Buick et al. 1995 ; Iizuka et al. 2006 ; Stern and Scholl2010 ; Adam et al. 2012). Supplementary evidence for exposed lands may consist of thick soils developed on these ancient surfaces (Buick et al. 1995 ; Hoffman 1995 ; Ohmoto et al. 2007 ; Johnson et al. 2009 , 2010). The further growth of continents and their sedimentary cover, implying extensive intracratonic terrestrial settings that remained relatively stable (although still affected by erosion, sea level changes, tectonics, and volcanism), is reflected in the ample record of paleosols onward (see study approaches and examples by Jackson 1967 ; Gay and Grandstaff 1980 ; Holland1984 ; Aspler and Donaldson 1986 ; Grandstaff et al. 1986 ; Kimberley and Grandstaff 1986 ; Reimer1986; Retallack 1986b ; Farrow and Mossman 1988 ; Zbinden et al. 1988 ; Palmer et al. 1989 ; Holland 1992 ; Gall 1994 ; Macfarlane et al. 1994 ; Martini 1994 ; Retallack and Mindszenty 1994 ; Driese et al. 1995 ; Banerjee 1996 ; Ohmoto 1996 ; Prasad and Roscoe 1996 ; Gutzmer and Beukes1998 ; Thiry and Simon-Coincon 1999 ; Rye and Holland 2000 ; Watanabe et al. 2000 ; Retallack2001 and references therein; Yang and Holland 2003 ; Driese and Gordon-Medaris 2008 ; Pandit et al. 2008 ; Bandopadhyay et al. 2010). This record of old paleosols holds indirect proof of the early environmental conditions on Earth and early life on land.

Currently, the oldest and direct evidence of terrestrial life comes from ~2,900–2,700 Ma-old (see age determination of Witwatersrand deposits in Kositcin and Krapez 2004 ; Zhao et al. 2006), organic matter–rich paleosols (Watanabe et al. 2000), ephemeral ponds (Rye and Holland 2000) and alluvial sequences, some of them bearing microfossils (Hallbauer and van Warmelo 1974 ; Mossman et al.2008). Interestingly, their occurrence in such settings coincides with drastic changes in Earth's crustal configuration and the —perhaps abrupt— emplacement of large continental masses in the late Archean (Condie 2004 ; Eriksson and Martins-Neto 2004 ; Van Kranendonk 2004 and references therein; Hazen et al. 2012), a marked step in the oxygenation of the atmosphere (Kendall et al. 2010), and also with estimations of land colonization by microbes based on phylogenetic relationships (Battistuzzi et al. 2004). Although microbes could have colonized the land before this time, the Meso- to Neoarchean appears to be an important evolutionary time period for terrestrial microbial communities, perhaps linked to supercontinent growth (Santosh 2010) and the emergence of potential new habitats.

Later in time, the amount of organic matter–rich and possibly "biologically weathered" paleosols (Ohmoto 1996 ; Beukes et al. 2002 ; Driese and Gordon-Medaris 2008), terrestrial sedimentary structures of presumed biotic origin (Hupe 1952 ; Lannerbro 1954 ; Voigt 1972 ; Eriksson et al. 2000; Prave 2002), and microfossils themselves (Cloud and Germs 1971 ; McConnell 1974 ; Horodyski and Knauth 1994 : Strother et al. 2011) drastically increased throughout the Proterozoic. Likewise, marine microfossils display increasing biological developments and adaptations (Knoll et al. 2006). Biotic diversity and abundance become even greater from the Neoproterozoic-Phanerozoic transition to the recent (see Zhuravlev and Riding 2001 ; Xiao and Kaufman 2006 ; Gaucher et al. 2010). This timeline suggests a rapid and global development of life on Earth, with life forms adapted to live on the land more than 2,000 Ma before the earliest fossil record of land plants (Heckman et al. 2001; Gensel 2008). Important events in this chronology are depicted in Figure 1.

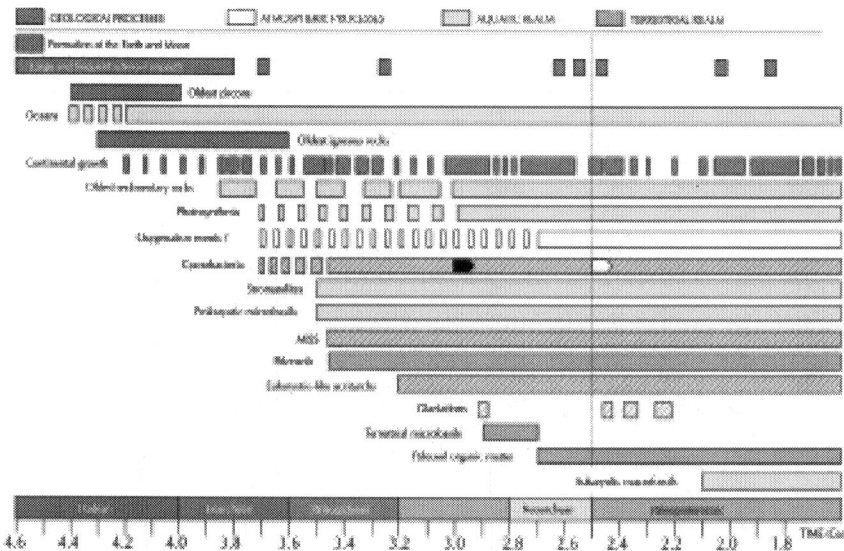

Figure 1: Suggested chronology of geological, atmospheric, and biological events during the Hadean, Archean, and Paleoproterozoic eons. Geological events were compiled from references within Canup and Righter (2000), Eriksson et al. (2004), and Van Kranendonk et al. (2007a,2007b). Asteroidal impact history is by Glikson (2007). Emplacement of the oceans is by Nutman (2006). The dashed line at 2,500 Ma marks an abrupt change into an oxygenated atmosphere (see Kendall et al. 2010), although oxygen may have started accumulating beforehand (Ohmoto 2004). Oxygenation events are considered to have occurred in pulses of unknown magnitude and duration and are correlated with a putative origin of oxygenic photosynthesis by cyanobacteria, based on a theoretical timing estimated by Lazcano and Miller (1994) and considering the emergence of life at 3,800 Ma. On the line of cyanobacteria, alternative origins are indicated with a black (Schirrmeister et al. 2013) or a white mark (Kirschvink and Kopp 2008). The MISS (Microbially Induced Sedimentary Structures; see Noffke et al. 2001), cyanobacteria, and the unicellular eukaryotic-like acritarchs are represented within both aquatic and terrestrial realms. Glaciations are from Hoffman and Schrag (2002), shown in combined colors to represent atmospheric (climatic) and hydrological (ice formation) processes. Other biological evolutionary steps and paleosol-related data are from Hallbauer and van Warmelo (1974), Holland (1984), Schopf (1983), Schopf and Klein (1992), Han and Runnegar (1992), Golubic and Seong-Joo (1999), Rye and Holland (2000), Retallack (2001), Westall et al. (2006a, 2006b), Johnson et al. (2009 ,2010), El Albani et al. (2010), Javaux et al. (2010), and Noffke (2010).

Functioning of Primitive Terrestrial Ecosystems and Cyanobacteria

A conceptualization of the functioning of the ancient terrestrial biosphere necessarily requires a general understanding of modern, analog microbial communities to evaluate their living requirements, diversity, physiology, and environmental impact, and to characterize any potential biosignature that could be used to recognize them in the rocks. Modern terrestrial microbial communities are found worldwide and in a great variety of local conditions, in surface (solid rock, regolith) and subsurface (caves, groundwater, deep ground) environments (although the latter could be considered aquatic by some). However, it is unclear which one is more productive in terms of biomass (Pace 1997) and what metabolisms have dominated those systems—and to what extent—over geologic time scales (Sleep and Bird 2007).

An understanding of the biology and distribution of modern microbes, which are ubiquitous in today's Earth's biosphere (Figure 2), seems essential for an understanding of their ancient counterparts and their impact on early terrestrial ecosystems. Estimates of the genetic diversity and biomass distribution in drastically different environments (e.g., Garcia-Pichel et al. 2003 ; Lozupone and Knight 2007 ; Nemergut et al. 2011) depict the ample range of strategies that terrestrial organisms, particularly primary producers, have developed for living on the land. Oxygenic photoautotrophy seems to be a particularly important capability of terrestrial organisms, simply because their energy source (light), reductant power (water), and carbon source (CO_2) are readily available in these environments. In comparison, other primary producers (e.g., chemolithotrophs) are restricted to aqueous environments because they require soluble sources of reductants (e.g., H_2, Fe^{2+}, H_2S, HS^-) and exergonic reactions to maintain their metabolism (White 2000). Besides being restricted to aquatic environments, chemolithotrophs are also less energy-efficient than oxygenic photoautotrophs (DesMarais 2000 ; Madigan et al. 2003 ; Konhauser 2007), and less likely dominant in subaerial environments.

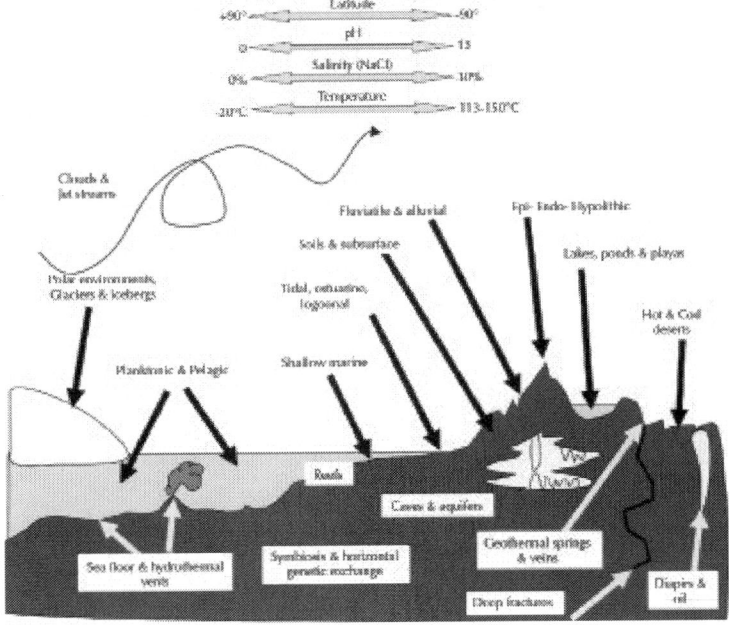

Figure 2: Variety of environments known to be inhabited by microbes. Ranges of environmental fluctuations (latitude, pH, salinity, and temperature) where microbes can be found are also shown (see Madigan and Marrs 1997; Madigan et al. 2003; Konhauser 2007). Direct and indirect interactions between different microbial communities, as well as symbiotic associations, coevolution, and horizontal genetic exchange (e.g., Davison 1999; Gogarten et al. 2007) are assumed to drive and have driven ecological processes and adaptations to habitat variability.

Cyanobacteria have been the only organisms that developed special pigments and enzymatic capabilities for using water as a source of electrons. This process has allowed them to live outside the water in any suitable environment, even where water might be a limiting factor, such as deserts (e.g., Potts and Friedmann 1981). Oxygenic photosynthesis also contributed to the oxidation of the atmosphere (both by sequestering CO_2 and by producing O_2), a global and ongoing process with profound geochemical, atmospheric, hydrological, and biological implications (e.g., Rosing et al. 2006; Och and Shields-Zhou 2012; Pufahl and Hiatt 2012). Cyanobacteria and other prokaryotes, can also fix gaseous nitrogen, which seems of great advantage for an independence from dissolved N species, such as NH_4 and NO_3 (Glass et al. 2009). The appearance of cyanobacterial akinetes (for N_2 fixation) in the Paleoproterozoic (Tomitani et al. 2006) attests to this early adaptation. For organisms on the land, a limiting nutrient, such as P, can be supplied by dust deposition (Kennedy et al. 1998; Reynolds et al. 2001 ; McTainsh and Strong 2007), which may be an alternative process for replenishment of nutrient loss by runoff and leaching in such environments (e.g. Beraldi-Campesi et al. 2009); S can also be acquired from minerals, aerosols, and as gaseous sources, likely present in the early atmosphere (Holland 1984). Thus, the nutritional requirements for oxygenic, photoautotrophic, primary producers seem not to have been a limiting factor for the colonization of the land. This idea has also been discussed in light of physiological and genetic characteristics of terrestrial microbes (Battistuzzi et al. 2004 ; Battistuzzi and Hedges 2009). Yet, an earlier chemotrophic way of life also needs to be considered (Shen and Buick 2004 ; Sleep and Bird 2007).

Particularly for the early terrestrial biota, a minimum set of adaptations to live subaerially must have included protection against radiation and desiccation effects. Adaptations such as thick polymeric sheaths with hygroscopic capacity against desiccation, efficient DNA repair mechanisms to restore metabolic activities as soon as water is available, and the production of UV-shielding pigments are certainly successful strategies displayed by terrestrial cyanobacteria (Shephard 1987 ; Garcia-Pichel 1998 ; Yasui and McCready 1998 ; Potts 1999 ; Sinha and Hader 2002 ; Singh et al. 2010). Refined degrees of adaptation of terrestrial organisms include sunscreens that once placed within the extracellular sheaths, passively protect against UV radiation, even when the cell is dormant or dehydrated (Garcia-Pichel and Castenholz 1991 ; Gao and Garcia-Pichel 2011). Some of these strategies for living on the land likely evolved early and are partially displayed by microfossils (e.g., thick sheaths), which are sometimes associated with evaporitic sediments, in keeping with subaerial exposure (Schopf 1968 ; Hofmann 1976 ; Golubic and Campbell 1979 ; Awramik et al. 1983).

Modern cyanobacteria-driven communities can be found in any terrestrial environment (~30% of modern Earth's surface area). They include endolithic communities (Friedmann 1980 ; Sun and Friedmann 1999 ; Büdel et al. 2004) and cryptogamic covers (CGC) on rocks and soils (Belnap and Lange 2001 ; Elbert et al. 2012). The latter have been shown to be very complex and dynamic and contain many distinct functional groups of prokaryotes and eukaryotes, ranging from primary producers to decomposers of specific materials, and grazers (Fritsch 1922 ; Fletcher and Martin 1948; Campbell 1979 ; Bamforth 1984 , 2004 ; Garcia-Pichel et al. 2001 ; Nagy et al. 2005 ; Tirkey and Adhikary 2005 ; Chanal et al. 2006 ; Reddy and Garcia-Pichel 2006 ; Bates and Garcia-Pichel 2009 ; Neher et al. 2009 ; Meadow and Zabinski 2012 Bates et al. 2013). This diversity is variable based on local environmental conditions, but all CGC—albeit with few exceptions (e.g., Hoppert et al. 2004)—have in common the presence of cyanobacteria. This speaks for the evolutionary success that cyanobacteria have had over other microbes throughout time.

Although fossil analogs of CGC have been discovered in ancient sediments (Simpson et al. 2010 ; Beraldi-Campesi et al. 2011 ; Retallack 2009 , 2011 ; Sheldon 2012), it is unknown what their microbial composition might have been. However, morphological similarities between modern and fossil counterparts are remarkable (Schieber et al. 2007 ; Noffke 2010). Morphological resemblance between fossils and recent analogs suggests that cyanobacteria are indeed a very old group of bacteria (see Golubic and Seong-Joo 1999) and that at least some morphological traits have been maintained over time (Golubic and Hofmann 1976 ; Golubic and Campbell 1979; Schopf 1992b). Moreover, as cyanobacteria are such an old group and are so well adapted to colonizing unstable sediments (Booth 1941 ; Campbell et al. 1989 ; Mazor et al. 1996 ; Belnap and Gillette 1998 ; Malam Issa et al. 2001 ; Hu et al. 2002 ; Garcia-Pichel and Wojciechowski 2009), even where available water is scarce and there is considerable UV radiation (Fleming and Castenholz 2007 ; Giordanino et al. 2011), it is likely that they were also primordial components on land surfaces (Campbell 1979) and influenced the formation of sedimentary biostructures and textures represented in fossil examples (e.g., Prave 2002 ; Schieber et al. 2007 ; Noffke, 2011). The antiquity of cyanobacteria has been also estimated by molecular clock analyses of genomic distances to be ~3,000 Ma (Battistuzzi and Hedges 2009 ; Schirrmeister et al. 2013), which more-or-less coincides with the age of the oldest terrestrial microfossils (Mossman et al. 2008). This timing, however, may vary depending on the calibration points used for constructing phylogenies and the extent of horizontal gene transfer. Lastly, the capacity of chlorophyll *a* to absorb higher photonic energies to split water in comparison to other photosynthetic bacteriochlorophylls (Xiong et al. 2000) may be the result of selective pressures to use the shorter wavelengths that reached the Precambrian surface where cyanobacteria had to dwell, a capacity not seen in purple or green phototrophic bacteria that use less energetic wavelengths in submerged/shielded habitats. Thus, from a multi-angular perspective cyanobacteria seem the perfect candidates for the colonization of the earliest lands.

As mentioned above, most CGC have in common the presence of filamentous cyanobacteria. One property of these morphotypes is that they can glide through interstitial spaces using hollow hygroscopic sheaths of mucilage as trails, to shield themselves against radiation, to find their optimum light regime, or to track water (Garcia-Pichel and Pringault 2001). The nature of the filamentous members of these communities also provides more surface area and tension for fastening and binding disaggregated particles (Garcia-Pichel and Wojciechowski 2009). Polysaccharides secreted extracellularly provide additional cementing force to the entire organo-mineral framework, resulting in the formation of a (crust/mat) stable microenvironment. The intrinsic characteristic of filamentous microbes to form cohesive layers at sedimentary surfaces is also known to substantially decrease wind and water erosion in modern arid and semiarid areas of the world (Belnap and Gillette1998 ; Belnap and Lange 2001). Although some erosive forces may surpass the tear resistance of CGC in high-energy systems (e.g., Corcoran and Mueller 2004), this property of microbes has been invoked to explain the stability of thick, Precambrian siliciclastic sedimentary sequences (Dott 2003) and the soft deformation properties of microbial mat-like structures (see references in Schieber et al.2007). This is an important property of microbes for the functioning of siliciclastic ecosystems, and together with the presence of mature and organic-rich soils and microfossils in old Proterozoic strata (see references above), suggests that abundant "cryptogamic" covers were present on Precambrian lands, similar to those covering polar and arid areas of the world today. The addition of new members to these communities over time (most importantly algae and fungi, but also grazers) is expected and may be used to explain increasing weathering rates of the continents (Kennedy et al. 2006) and abrupt changes in the global balance of the C cycle in the Neoproterozoic (Knauth and Kennedy2009).

Other Microbial Components

Judging from the rapid achievement of diversity and distribution of early microbial biota and from microbial successions seen in modern "barren" lands (e.g., Sigler et al. 2002 ; Schmidt et al. 2008 ; Fierer et al. 2010), it is expected that heterotrophic organisms were also part of land communities, as they seem to be an inevitable component in this type of consortia. Under this perspective, primitive microbial ecosystems cannot be understood as composed only of autotrophic primary producers, but also a myriad of other microbes finding their niche within such pre-existent microenvironments. For example, actinobacteria in modern CGC not only degrade large quantities of organic exudates from cyanobacteria, a process which influences the C cycle, but they also seem to be structural components of these sedimentary biostructures (e.g., Reddy and Garcia-Pichel 2006). The same applies to other bacteria (e.g., Bacteroidetes and Proteobacteria) that secrete large quantities of mucopolysaccharides, which aid in gluing soil particles together and may also have a critical role in the hydraulic conductivity of the surface substrate (Rossi et al. 2012). One of the most important eukaryotic component of of modern CGC are fungi, which must have played a key role in the colonization and weathering of bare rocks in the past (with symbiont cyanobacteria or algae), as well as in the successive development of vascular plants on the land (Smith and Read 2008) and in a radical change toward more "modern" terrestrial ecosystems (Blackwell 2000 ; Heckman et al. 2001 ; Taylor et al. 2009 . See also Gadd 2006).

Although the timing of the origin of these organisms is unknown, terrestrial microbes can certainly drive important chemical transformations in soils (Keller and Wood 1993 ; Schwartzman and Volk1989 ; Chenu and Stotzky 2002 ; Ehrlich 2002 ; Chorover et al. 2007) and endolithic habitats (Konhauser et al. 1994 ; Sun and Friedmann 1999 ; Büdel et al. 2004 ; Omelon et al. 2006) that may have operated in the past. These include affecting the reactivity of mineral surfaces with secreted metabolites (Geesey and Jang 1990 ; Welch et al. 1999), changing the pH and redox potential of the microenvironment (Bennett et al. 2001), or secreting metal ligands and other organic complexes that react with solutes and minerals (Keller and Wood 1993 ; Schwartzman and Volk 1989 ; Barker et al.1998 ; Welch et al. 1999 ; Bennett et al. 2001). These mechanisms seem to play a fundamental role in biogeochemistry (weathering, clay formation, nutrient bioavailability, metal concentration and bioavailability, mineral formation or transformation, etc.), and their effects may also be used to trace microbial geochemical biosignatures in the rock record (Beraldi-Campesi et al. 2009). Additionally, the process of soil formation and maturation is usually understood as aided by biology (Keller and Wood 1993 ; Schwartzman and Volk 1989 ; Brady and Weil 2008) and differentiated from abiotic regolith development, and involves a critical step prior to plant and animal colonization of the land. All these characteristics displayed by modern CGC could be expected from ancient analog communities, although with variations in the occurrence and magnitude dictated by their limiting factors.

Dust

The mechanism of dust formation, transport, and deposition reflects one important aspect of the functioning of terrestrial ecosystems because dust can only be formed on the land and because microbes (along with water adhesion and neo-cementation of particles with salts and clays) can stabilize fine dust particles through trapping and binding (e.g., Dong et al. 1987 ; Liu et al. 1994 ; Williams et al. 1995 ; Belnap and Gillette 1998 ; Hu et al. 2002). Thus, dust production can potentially be regulated by microbes (and other encrusting processes) depending on their degree of development. The more developed, the less dust production.

Dust is an important carrier of nutrients, and its retention on the ground might influence the budget and delivery of those nutrients locally or to other distant ecosystems, such as happens in modern marine environments via deposition of huge loads of dust (Jickells et al. 2005). The capacity of microbes to trap and bind particles has been demonstrated for numerous underwater and subaerial environments (Gunatilaka 1975 ; Zhang 1992 ; Takeuchi et al. 2001 ; Altermann 2008 ; Gradzinski et al. 2010 ; Williams et al. 2012). If microbes were responsible for much of the global dust capture, retention, and lixiviation on the early continents, recycling effects may have had profound implications for the evolution of global ecosystems through geologic time, as well as for important climatic processes, such as those rooted in atmospheric albedo variations (Harrison et al. 2001 ; Jickells et al. 2005 ; Lau et al. 2006).

Finally, dust is also a carrier of microbes and viruses (Abed et al. 2011 ; Al-Bader et al. 2012), which implies a means for biological dispersion that must have been operating continuously and over long distances in the past, amplifying the potential biogeography of biological entities over vast areas of the oceans and continents. Although the rate of survival and the success of foreign airborne mixed communities on aquatic environments, barren or already colonized surfaces is not known, it is plausible that such a mechanism was vital for the colonization of the early continents and the increase in ecological complexity and genetic exchange (e.g., Gogarten et al. 2007).

Underground Realm

The underground realm (geothermal veins, aquifers, soil subsurface, all types of caves) should also be considered potential continental habitats for early terrestrial life, as life is abundant there today (e.g., Ghiorse and Wilson 1988 ; Barton and Northup 2007 ; Engel 2010). The Precambrian record of caves (e.g., karstic environments) or underground aquifers (detected through nodules and concretions in the rocks) is far less known than the typical shallow marine or lacustrine environments (see examples of karstic and underground environments in Nicholas and Bildgen 1979 ; Schau and Henderson 1983 ; Glover and Kah 2006 ; Skotnicki and Knauth 2007 ; Rasmussen et al. 2009). Nevertheless, these environments must have existed throughout Earth's history, and thus, terrestrial biotas could have adapted to live in those conditions back in the Precambrian (Rasmussen et al.2009).

In contrast to the most typical subaerial, light-driven primary producer communities, underground microbes require a chemosynthetic metabolism for primary productivity, perhaps relying on the oxidation of sulfur and iron compounds to support growth and continuity, as these are the main energy sources in such environments (Sarbu et al. 1996 ; Chen et al. 2009 ; Porter et al. 2009). Because these metabolic pathways are less energetic than photosynthesis, life underground is expected to have been slow-growing, less dynamic in terms of diversity and interactions, and more geographically contained than, for example, subaerial phototrophs. Nevertheless, early underground dwellers may have impacted the subsurface realm (cave formation, buried oil and dissolved organic matter consumption, methane production, etc.) and contributed to the neoformation and dissolution of minerals over the long term, as well as to the generation of gaseous byproducts (e.g., H_2S, CH_4, CO_2) that could be important for geochemical processes on the surface and ultimately for distant communities and global biogeochemical recycling. Moreover, this type of environments could have been better protected from drastic and global crisis than subaerial ones, and thus have functioned as living reservoirs that could later exploit surface environments.

A Note on Biosignatures

Imprints of life in rocks can be formed in various ways and can be recognized as long as the rocks are preserved and accessible. Although this "fossil" record decreases in outcrop abundance the older the rocks are (basically due to burial, erosion, subduction, and metamorphism), biosignatures can be found in sedimentary rocks (Schopf 1983 ; Schopf and Klein 1992 ; Schieber et al. 2007 ; Noffke2010), but also in igneous (Banerjee et al. 2006 ; Furnes et al. 2004, 2007b ; Fliegel et al. 2010) and metamorphic (Franz et al. 1991 ; Hanel et al. 1999 : Squire et al. 2006 ; Bernard et al. 2007 ; Schiffbauer et al. 2007 , 2012 ; Schiffbauer and Xiao 2009 ; Zang 2007) rocks of all ages.

Preservation will always be favored in underwater settings, especially if biological materials are buried quickly, the sediment is fine grained, and the conditions are overall reducing (anoxic). All of these factors promote rapid mineralization and replacement of biological materials (Farmer 1999 ; Zonneveld et al. 2010 ; Allison and Bottjer 2011 ; Lalonde et al. 2012) which can preserve the morphology and organic remains, although this does not mean that preservation always happens (Zonneveld et al. 2010and references therein). Unless protected, organic matter tends to degrade basically by photo-chemical degradation (if exposed to light), chemical bond breaking, biological recycling, mechanical maceration and dissolution. If body fossils are preserved, the lack of diagnostic morphologies for most bacteria and the possible existence of abiotic, microbe-like morphologies (e.g., García-Ruiz et al. 2002 , Garcia-Ruiz et al. 2003) make their determination a challenge. Yet, their presence in a suitable geological context and association with sedimentary biostructures may be used as criteria for biogenicity. Molecular biomarkers in hydrocarbons that can be correlated with extant organisms (e.g., Summons et al. 1999) also require careful confirmation of syngenicity for a correct interpretation (Rasmussen et al. 2008 ; Brocks 2011).

If limiting factors are at play, microbial communities may not develop sufficient biomass to leave behind a biosignature (either chemical, geochemical, mineralogical, or morphological). Water, for example, which is the most basic requirement for survival and reproduction, tends to be a limiting factor on the land compared to a permanent water body. If microbial growth is thus limited, the amount of cells and biomass that can be preserved decreases. Thus, organisms with access to unlimited water resources would be able to grow larger communities and have more possibilities for fossilization, in contrast to terrestrial microbes that depend on dew or rain for their survival and maintenance. For example, the thickness and cohesiveness of a marine-intertidal microbial mat (see Bauld 1981 ; Bauld et al. 1992) are greater than in a mature biological soil crust (Belnap and Lange2001), thus the latter will be less prone to fossilization than the marine one. Nevertheless, under favorable climates and landscapes, these too could be preserved (e.g., Prave 2002). Studies on biosignatures left behind by terrestrial microbial communities are needed for comparison against the yet-to-explore rock record.

CONCLUSIONS

As the Earth was evolving, gradual degassing and accumulation of liquid water at its surface differentiated aquatic and non-aquatic environments. Because terrestrial environments have always existed, it is possible that life evolved on the land (including in lakes, rivers, streams, and flooding areas) as early as aquatic life itself (see Retallack 1986aand references therein). In any case, living on the land must have required particular adaptations, such as the capability to acquire nutrients and energy sources outside the aquatic realm, the development of molecular repairing mechanisms, and protection against radiation and desiccation. These adaptations were certainly developed by cyanobacteria, a group with a very old biologic lineage and one of the most conspicuous and successful primary producers on Earth (e.g., Whitton and Potts 2000 ; Herrero and Flores 2008).

Direct and indirect evidence pointing to inhabited terrestrial environments by the Paleoarchean (Johnson et al. 2009 , 2010) and the following eras (Stüeken et al. 2012), along with substantive evidence of terrestrialization from the Neoarchean onward (Hallbauer and van Warmelo 1974 ; McConnell 1974 ; Horodyski and Knauth 1994 ; Gutzmer and Beukes 1998 ; Rye and Holland 2000 ; Watanabe et al. 2000 ; Prave 2002 ; Rasmussen et al. 2009), strongly implies that functional terrestrial ecosystems originated well back in the Precambrian. The implications for such colonization have not been completely understood, but the effects of microbial life on land processes that affect the atmosphere, the lithosphere, and the hydrosphere, are widely diverse and act at all scales. Two main consequences derived from the activities of land biota are the continuous oxygenation of the atmosphere (with consequences for the stratification of the oceans, the formation and maintenance of the ozone layer, and the precipitation of oxides, among others) and the weathering of the continents (Holland 1984 ; Catling et al. 2001 ; Stüeken et al. 2012), which indirectly and directly affect marine ecosystems. In contrast to marine biota, which indirectly affect terrestrial ecosystems through atmospheric processes (including gas composition and climate), the establishment of life on the land has an enormous significance for the evolution of the planet through time because gaseous byproducts, such as oxygen produced on the land would be released directly into the atmosphere and not dissolved in the oceans first. Once in the atmosphere oxygen would react with reduced species before starting to accumulate and produce a geochemical signature in marine sediments. Thus, land-based life could have been pivotal for the early oxygenation of the atmosphere, which later affected the oceans as well. A more direct influence of land-based communities over aquatic ones would be the production of dust, clays and leachates on the continents (Kennedy and Wagner 2011 and references therein), which would then be carried by rivers and wind into the oceans, and thus increasing the heterogeneity of materials and solutes delivered into oceanic ecosystems and having either beneficial or detrimental consequences for marine life. Yet, an overall retention of sediments on the land via microbial stabilization would be expected for detritic sediments in places with well-developed cryptogamic covers. Finally, it is expected that the time span from the inception of microbial land-based life to the evolution of the first plant ecosystems took long enough (2,000–2,500 Ma) for coastal and inland settings to be transformed into organic- and nutrient-rich substrates that could later be exploited by more evolved communities and organisms toward the Neoproterozoic-Phanerozoic transition.

In general, the logical transition from cyanobacteria (and other bacteria and archea), to algae (and protists and fungi), to non-vascular plants, to vascular plants, may still be valid, but the timing of those evolutionary steps needs to be updated with the latest pertinent information available. The notion that the land was virtually sterile in the Precambrian (e.g., Bambach 1999 ; Blackwell 2000 ; Corcoran and Mueller 2004 ; Nesbitt and Young 2004 ; Gensel 2008) underestimates the impact that microbes could have had on the biosphere. More importantly, the idea the land was first colonized by plants and that they formed the earliest terrestrial ecosystems should be abandoned completely. That is not to say that the advent of plants in the Phanerozoic did not have strikingly enhanced effects on continental weathering, soil formation, and oxygenation of the atmosphere (Labandeira 2005 ; Taylor et al. 2009), but neglecting the existence of microbial, Paleoarchean-to-recent terrestrial ecosystems would impede a realistic understanding of the evolution of the biosphere and its influence on the geo-atmo-hydrosphere over time.

ACKNOWLEDGMENTS

I am grateful to Kathleen E. Pigg, Anthony R. Prave, Gregory J. Retallack, Nora Noffke, Fernando Ortega Gutiérrez, Dominic Papineau, Marcela Martínez Millán, and Kelaine Ravdin for their important comments and improvements to this paper. I also thank the editors of Springer and Bettina Weber and Jayne Belnap for organizing and editing this special volume. I also thank the people from SIOV (Seminario Interdisciplinario del Origen de la Vida) at UNAM for fruitful discussions on critical aspects of this topic.

REFERENCES

1. Abed RMM, Ramette A, Hübner V, De Dekker P, de Beer D (2011) Microbial diversity of eolian dust sources from saline lake sediments and biological soil crusts in arid Southern Australia. FEMS Microbiol Ecol 80(2):294-304

2. Adam J, Rushmer T, O'Neil J, Francis D (2012) Hadean greenstones from the Nuvvuagittuq fold belt and the origin of the Earth's early continental crust. Geology 40:363-366

3. Al-Bader D, Eliyas M, Rayan R, Radwan S (2012) Air-dust-borne associations of phototrophic and hydrocarbon-utilizing microorganisms: promising consortia in volatile hydrocarbon bioremediation. Environ Sci Pollut R 19(9):3997-4005

4. Aller JY, Aller RC, Kemp PF, Chistoserdov AY, Madrid VM (2010) Fluidized muds: a novel setting for the generation of biosphere diversity through geologic time. Geobiology 8:169-178

5. Allison PA, Bottjer DJ (2011) Taphonomy: bias and process through time. In: Allison PA, Bottjer DJ (eds) Taphonomy: process and bias through time. Topics in geobiology, vol. 32, New York: Springer. pp 1-17

6. Allwood AC, Walter MR, Kamber BS, Marshall CP, Burch IW (2006) Stromatolite reef from the Early Archaean Era of Australia. Nature 441:714-718

7. Altermann W (2008) Accretion, trapping and binding of sediment in Archean stromatolites—morphological expression of the antiquity of life. In: Botta O, Bada J, Gómez EJ, Javaux E, Selsis F, Summons R (eds) Strategies of life detection. Space Sciences Series of ISSI, vol. 25, New York: Springer. pp 55-79

8. Altermann W, Schopf JW (1995) Microfossils from the Neoarchean Campbell Group, Griqualand West Sequence of the Transvaal Supergroup, and their paleoenvironmental and evolutionary implications. Precambrian Res 75:65-90

9. Aspler LB, Donaldson JA (1986) Paleoclimatology of Nonacho Basin (Early Proterozoic), Northwest Territories, Canada. Palaeogeogr Palaeoclimateol Palaeoecol 56:17-34

10. Awramik S (1992) The oldest records of photosynthesis. Photosynth Res 33:75-89

11. Awramik SM, Buchheim HP (2009) A giant, Late Archean lake system: the Meentheena Member (Tumbiana Formation; Fortescue Group), Western Australia. Precambrian Res 174:215-240

12. Awramik SM, Schopf JW, Walter MR (1983) Filamentous fossil bacteria from the Archean of Western Australia. Precambrian Res 20:357-374

13. Bambach RK (1999) Energetics in the global marine fauna: a connection between terrestrial diversification and change in the marine biosphere. Geobios 32(2):131-144

14. Bamforth SS (1984) Microbial distributions in Arizona deserts and woodlands. Soil Biol Biochem 16(2):133-137

15. Bamforth SS (2004) Water film fauna of microbiotic crusts of a warm desert. J Arid Environ 56:413-423

16. Bandopadhyay PC, Eriksson PG, Roberts RJ (2010) A vertic paleosol at the Archean-Proterozoic contact from the Singhbhum-Orissa craton, eastern India. Precambrian Res 177(3–4):277-290

17. Banerjee DM (1996) A lower Proterozoic paleosol at BGC–Aravalli boundary in south-central Rajasthan. India J Geol Soc 48:277-288

18. Banerjee NR, Furnes H, Muehlenbachs K, Staudigel H, de Wit MJ (2006) Preservation of ca. 3.4–3.5 Ga microbial biomarkers in pillow lavas and hyaloclastites from the Barberton Greenstone Belt, South Africa. Earth Planet Sci Lett 241:707-722

19. Barker WW, Welch SA, Chu S, Banfield JF (1998) Experimental observations of the effects of bacteria on aluminosilicate weathering. Am Mineral 83:1551-1563

20. Baross JA, Hoffman SE (1985) Submarine hydrothermal vents and associated gradient environments as sites for the origin and evolution of life. Orig Life 15:327-345

21. Barton HA, Northup DE (2007) Geomicrobiology in cave environments: past, current and future perspectives. J Cave Karst Stud 69:163-178

22. Bates ST, Garcia-Pichel F (2009) A culture-independent study of free-living fungi in biological soil crusts of the Colorado Plateau: their diversity and relative contribution to microbial biomass. Environ Microbiol 11(1):56-67

23. Bates ST, Clemente JC, Flores GE, Walters WA, Wegener-Parfrey L, Knight R, Fierer N (2013) Global biogeography of highly diverse protistan communities in soil. ISME Jour 7:652-659

24. Battistuzzi FU, Hedges SB (2009) A major clade of prokaryotes with ancient adaptations to life on land. Mol Biol Evol 26:335-344

25. Battistuzzi FU, Feijao A, Hedges SB (2004) A genomic timescale of prokaryote evolution: insights into the origin of methanogenesis, phototrophy, and the colonization of land. BMC Evol Biol 4(44):1-14

26. Bauld J (1981) Geobiological role of cyanobacterial mats in sedimentary environments: production and preservation of organic matter. BMR J Aust Geol Geophys 6:307-317

27. Bauld J, D'Amelio E, Farmer JD (1992) Modern microbial mats. In: Schopf JW, Klein C (eds) The Proterozoic biosphere, New York: Cambridge University Press. pp 261-269

28. Behrensmeyer AK, Damuth JD, DiMichele WA, Potts R (1992) Terrestrial ecosystems through time: evolutionary paleoecology of terrestrial plants and animals. Chicago Il: University of Chicago Press.

29. Belnap J, Gillette DA (1998) Vulnerability of desert biological soil crusts to wind erosion: the influences of crust development, soil texture, and disturbance. J Arid Environ 39:133-142

30. Belnap J, Lange OL (2001) Biological soil crusts: structure, function, and management. Ecological Studies Series, vol. 150. Berlin: Springer.

31. Bennett PC, Rogers JR, Choi WJ (2001) Silicates, silicate weathering, and microbial ecology. Geomicrobiol J 18:3-19

32. Beraldi-Campesi H, Hartnett HE, Anbar A, Gordon GW, Garcia-Pichel F (2009) Effect of biological soil crusts on soil elemental concentrations: implications for biogeochemistry and as traceable biosignatures of ancient life on land. Geobiology 7:348-359

33. Beraldi-Campesi H, Farmer JD, Garica-Pichel F (2011) Evidence for Mesoproterozoic life on land and its modern counterpart in arid soils. Proceedings of the GSA Annual Meeting, Minneapolis, 9–12 October 2011.

34. Bernard S, Benzerara K, Beyssac O, Menguy N, Guyot F, Brown GE Jr, Goffe B (2007) Exceptional preservation of fossil plant spores in high-pressure metamorphic rocks. Earth Planet Sc Lett 262:257-272

35. Beukes NJ, Dorland H, Gutzmer J, Nedachi M, Ohmoto H (2002) Tropical laterites, life on land, and the history of atmospheric oxygen in the Paleoproterozoic. Geology 30:491-494

36. Blackwell M (2000) Terrestrial life: fungal from the start? Science 289(5486):1884-1885

37. Blank CE (2009) Not so old Archaea—the antiquity of biogeochemical processes in the archaeal domain of life. Geobiology 7(5):495-514

38.   Booth WE (1941) Algae as pioneers in plant succession and their importance in erosion control. Ecology 22:38-46

39.   Bouvier W (2010) The age of the Solar System redefined. Nat Geosci 3:637-641

40.   Boyet M, Carlson RW (2005) ^{142}Nd evidence for early (>4.53 Ga) global differentiation of the silicate Earth. Science 309:576-581

41.   Brady NC, Weil RR (2008) The nature and properties of soils. Upper Saddle River, NJ: Pearson-Prentice Hall.

42.   Brocks JJ (2011) Millimeter-scale concentration gradients of hydrocarbons in Archean shales: live-oil escape or fingerprint of contamination? Geochim Cosmochim Acta 75(11):3196-3213

43.   Büdel B, Weber B, KuhlM PH, Sultemeyer D, Wessels D (2004) Reshaping of sandstone surfaces by cryptoendolithic cyanobacteria: bioalkalisation causes chemical weathering in arid landscapes. Geobiology 2:261-268

44.   Buick R (2008) When did oxygenic photosynthesis evolve? Philos Trans R Soc Lond B Biol Sci 363(1504):2731-2743

45.   Buick R (2010) Early life: ancient acritarchs. Nature 463(7283):885-886

46.   Buick R, Thornett JR, McNaughton NJ, Smith JB, Barley ME, Savage M (1995) Record of emergent continental crust 3.5 billion years ago in the Pilbara Craton of Australia. Nature 375:574-577

47.   Campbell SE (1979) Soil stabilization by a prokaryotic desert crust: implications for Precambrian land biota. Orig Life 9:335-348

48.   Campbell SE, Seeler J, Golubic S (1989) Desert crust formation and soil stabilization. In: J. Skujins (ed) Uses of microbiological processes in arid lands for desertification control and increased productivity (UNEP). Arid Soil Res Rehab 3:217-228

49.   Canup RM, Righter K (eds) (2000) Origin of the Earth and moon Tucson, AZ: University of Arizona Press.

50.   Catling DC, Zahnle KJ, McKay CP (2001) Biogenic methane, hydrogen escape, and the irreversible oxidation of early Earth. Science 293:839-843

51.   Cavosie AJ, Valley JW, Wilde S (2007) The oldest terrestrial mineral record: a review of 4400 to 3900 Ma detrital zircons from Jack Hills, Western Australia. In: Van Kranendonk MJ, Smithies RH,

BennettV (eds) Earth's oldest rocks. developments in Precambrian geology, Series 15, Amsterdam: Elsevier. pp 91-111

52. Chanal A, Chapon V, Benzerara K, Barakat M, Christen R, Achouak W, Barras F, Heulin T (2006) The desert of Tataouine: an extreme environment that hosts a wide diversity of microorganisms and radiotolerant bacteria. Environ Microbiol 8(3):514-525

53. Chen Y, Wu L, Boden R, Hillebrand A, Kumaresan D, Moussard H, Baciu M, Lu Y, Murrell JC (2009) Life without light: microbial diversity and evidence of sulfur- and ammonium-based chemolithotrophy in Movile Cave. ISME J 3:1093-1104

54. Chenu C, Stotzky G (2002) Interactions between microorganisms and soil particles: an overview. In: Huang PM, Bollag JM, Senesi N (eds) Interactions between soil particles and microorganisms: impact on the terrestrial ecosystem, Chichester: John Wiley and Sons. pp 4-40

55. Chorover J, Kretzschmar R, Garcia-Pichel F, Sparks DL (2007) Soil biogeochemical processes within the Critical Zone. Elements 3:321-326

56. Cloud P, Germs A (1971) New pre-paleozoic nannofossils from the Stoer formation (Torridonian), Northwest Scotland. Geol Soc Am Bull 82:3469-3474

57. Condie KC (2004) Precambrian superplume events. In: Eriksson PG, Altermann W, Nelson DR, Mueller WU, Catuneanu O (eds) The Precambrian Earth: tempos and events. Developments in Precambrian geology, vol. 12, Amsterdam: Elsevier. pp 163-173

58. Corcoran PL, Mueller WU (2004) Aggressive Archaean weathering. In: Eriksson PG, Altermann W, Nelson DR, Mueller WU, Catuneanu O (eds) The Precambrian Earth: tempos and events. Developments in Precambrian geology, vol. 12, Amsterdam: Elsevier. pp 494-504

59. Dauphas N, van Zuilen M, Wadhwa M, Davis AM, Martey B, Janney PE (2004) Clues from Fe isotope variations on the origin of early Archean BIFs from Greenland. Science 302:2077-2080

60. David LA, Alm EJ (2011) Rapid evolutionary innovation during an Archaean genetic expansion. Nature 469(7328):93-96

61. Davison J (1999) Genetic exchange between bacteria in the environment. Plasmid 42(2):73-91

62. Deming D (2002) Origin of the ocean and continents: a unified theory of the earth. Int Geol Rev 44:137-152

63. DesMarais DJ (2000) When did photosynthesis emerge on Earth? Science 289:1703-1705

64. DiMichele WA, Hook RW (1992) Paleozoic terrestrial ecosystems. In: Behrensmeyer AK, Damuth JD, DiMichele WA, Potts R, Suess HD, Wing SL (eds) Terrestrial ecosystems through time, Chicago: Chicago University Press. pp 205-325

65. Dong GR, Li CZ, Jin T, Gao SY, Wu D (1987) Some results on soil wind-tunnel imitating experiment. Chinese Sci Bull 32:297-301

66. Dott RH Jr (2003) The importance of eolian abrasion in supermature quartz sandstones and the paradox of weathering on vegetation-free landscapes. J Geol 111:387-405

67. Driese SG, Gordon-Medaris L Jr (2008) Evidence for biological and hydrological controls on the development of a Paleoproterozoic paleoweathering profile in the Baraboo Range, Wisconsin, USA. J Sediment Res 78:443-457

68. Driese SG, Simpson EL, Eriksson KA (1995) Redoximorphic paleosols in alluvial and lacustrine deposits, 1.8 Ga lochness formation, Mount Isa, Australia: pedogenic processes and implications for paleoclimate. J Sediment Res A65:675-689

69. Ehrlich HL (2002) Geomicrobiology. New York, NY: Marcel Dekker.

70. El Albani A, Bengtson S, Canfield DE, Bekker A, Macchiarelli R, Mazurier A, Hammarlund EU, Boulvais P, Dupuy JJ, Fontaine C, Fürsich FT, Gauthier-Lafaye F (2010) Large colonial organisms with coordinated growth in oxygenated environments 2.1 Gyr ago. Nature 466(7302):100-104

71. Elbert W, Weber B, Burrows S, Steinkamp J, Büdel B, Andreae MO, Pöschl U (2012) Contribution of cryptogamic covers to the global cycles of carbon and nitrogen. Nat Geosci 5:459-462

72. Engel AS (2010) Microbial diversity of cave ecosystems. In: Barton L, Mandl M, Loy A (eds) Geomicrobiology: molecular and environmental perspectives, New York: Springer. pp 219-238

73. Eriksson PG, Martins-Neto MA (2004) Commentary. In: The Precambrian Earth: tempos and events. Developments in Precambrian geology, vol. 12. Amsterdam: Elsevier. pp 677-680

74. Eriksson PG, Simpson EL, Eriksson KA, Bumby AJ, Steyn GL, Sarkar S (2000) Muddy roll-up structures in siliciclastic interdune beds of the c. 1.8 Ga Waterberg Group, South Africa. PALAIOS 15:177-183

75. Eriksson PG, Altermann W, Nelson DR, Mueller WU, Catuneanu O (2004) The Precambrian Earth: tempos and events. Developments in Precambrian geology, vol. 12. Amsterdam: Elsevier.

76. Farmer J (1999) Taphonomic modes in microbial fossilization. In: Space Studies Board (ed) Size limits of very small microorganisms. Washington DC: National Research Council, National Academy Press. pp 94-102

77. Farrow CE, Mossman DJ (1988) Geology of Precambrian paleosols at the base of the Huronian supergroup, Elliot Lake, Ontario, Canada. Precambrian Res 42:107-139

78. Fedo CM, Whitehouse MJ (2002) Metasomatic origin of quartz-pyroxene rock, Akilia, Greenland, and implications for Earth's earliest life. Science 296:1448-1452

79. Fierer N, Nemergut D, Knight R, Craine JM (2010) Changes through time: integrating microorganisms into the study of succession. Res Microbiol 161:635-642

80. Fleming ED, Castenholz RW (2007) Effects of periodic desiccation on the synthesis of the UV-screening compound, scytonemin, in cyanobacteria. Environ Microbiol 9:1448-1455

81. Fletcher JE, Martin WP (1948) Some effects of algae and mold in the rain-crust of desert soils. Ecology 29(1):95-100

82. Fliegel D, Wirth R, Simonetti A, Furnes H, Staudigel H, Hanski E, Muehlenbachs (2010) Septate-tubular textures in 2.0-Ga pillow lavas from the Pechenga Greenstone Belt: a nano-spectroscopic approach to investigate their biogenicity. Geobiology 8:372-390

83. Franz G, Mosbrugger V, Menge R (1991) Carbo-Permian pteridophyll leaf fragments from an amphibolite facies basement, Tauern Window, Austria. Terra Nova 3:137-141

84. Friedmann EI (1980) Endolithic microbial life in hot and cold deserts. Orig Life 10:223-235

85. Fritsch FE (1922) The terrestrial alga. J Ecol 10(2):220-236

86. Furnes H, Banerjee NR, Muehlenbachs K, Staudigel H, de Wit M (2004) Early life recorded in Archean pillow lavas. Science 304:578-81

87. Furnes H, de Wit M, Staudigel H, Rosing M, Muehlenbachs K (2007) Vestige of Earth's oldest ophiolite. Science 315:1704-1707

88. Furnes H, Banerjee NR, Staudigel H, Muehlenbachs K, McLoughlin N, de Wit M, Van Kranendonk M (2007) Comparing petrographic signatures of bioalteration in recent to Mesoarchean pillow lavas: tracing subsurface life in oceanic igneous rocks. Precambrian Res 158:156-176

89. Gadd GM (ed) (2006) Fungi in biogeochemical cycles Cambridge, UK: Cambridge University Press.

90. Gall Q (1994) The Proterozoic Thelon paleosol, Northwest Territories, Canada. Precambrian Res 68:115-137

91. Gao Q, Garcia-Pichel F (2011) Microbial ultraviolet sunscreens. Nat Rev Microbiol 9:791-802

92. García Ruiz JM, Carnerup A, Christy AG, Welham NJ, Hyde ST (2002) Morphology: an ambiguous indicator of biogenicity. Astrobiology 2(2):353-369

93. Garcia-Pichel F (1998) Solar ultraviolet and the evolutionary history of cyanobacteria. Origins Life Evol. Biosphere 28:321-347

94. Garcia-Pichel F, Castenholz RW (1991) Characterization and biological implications of scytonemin, a cyanobacterial sheath pigment. J Phycol 27:395-409

95. Garcia-Pichel F, Pringault O (2001) Cyanobacteria track water in desert soils. Nature 413:380-381

96. Garcia-Pichel F, Wojciechowski MF (2009) The evolution of a capacity to build supra-cellular ropes enabled filamentous cyanobacteria to colonize highly erodible substrates. PLoS One 4(11):e7801

97. Garcia-Pichel F, Lopez-Cortes A, Nubel U (2001) Phylogenetic and morphological diversity of cyanobacteria in soil desert crusts from the Colorado Plateau. Appl Environ Microbiol 67:1902-1910

98. Garcia-Pichel F, Belnap J, Neuer S, Schanz F (2003) Estimates of global cyanobacterial biomass and its distribution. Algol Stud 109:213-227

99. Garcia-Ruiz JM, Hyde ST, Carnerup AM, Christy AG, Van Kranendonk MJ, Welham NJ (2003) Self-assembled silica-carbonate structures and detection of ancient microfossils. Science 302:1194-1197

100. Gaucher C, Sial AN, Halverson GP, Frimmel HE (2010) Neoproterozoic-Cambrian tectonics, global change and evolution: a focus on Southwestern Gondwana. Developments in Precambrian Geology, vol. 16. Amsterdam: Elsevier.

101. Gay AL, Grandstaff DE (1980) Chemistry and mineralogy of Precambrian paleosols at Elliot Lake, Ontario, Canada. Precambrian Res 12:349-373

102. Geesey G, Jang L (1990) Extracellular polymers for metal binding. In: Ehrlich HL, Brierley CL (eds) Microbial mineral recovery, New York: McGraw-Hill. pp 223-249

103. Gensel PG (2008) The earliest land plants. Annu Rev Ecol Evol Syst 39:459-477

104. Ghiorse W, Wilson J (1988) Microbial ecology of the terrestrial subsurface. In: Laskin A (ed) Advances in applied microbiology, New York: Academic. pp 107-172

105. Giordanino F, Strauch SM, Villafañe VE, Helbling EW (2011) Influence of temperature and UVR on photosynthesis and morphology of four species of cyanobacteria. J Photochem Photobiol B: Biol 103:68-77

106. Glass JB, Wolfe-Simon F, Anbar AD (2009) Coevolution of metal availability and nitrogen assimilation in cyanobacteria and algae. Geobiology 7:100-123

107. Glikson A (2007) Early Archean asteroid impacts on Earth: stratigraphic and isotopic age correlations and possible geodynamic consequences. In: Van Kranendonk MJ, Smithies RH, Bennett V (eds) Earth's oldest rocks. Developments in Precambrian geology, Series 15, Amsterdam: Elsevier. pp 1087-1103

108. Glover JF, Kah LC (2006) Speleothem deposits in a Proterozoic paleokarst, Mesoproterozoic dismal lakes group, Arctic Canada. Geological Society of America Abstracts with Programs 38(3):36

109. Gogarten JP, Fournier G, Zhaxybayeva O (2007) Gene transfer and the reconstruction of life's early history from genomic data.

In: Botta O, Bada JL, Gomez-Elvira J, Javaux E, Selsis F, Summons R (eds) Strategies of life detection. Space Science Series of ISSI, vol. 25, New York: Springer. pp 115-131

110. Golubic S, Campbell SE (1979) Analagous microbial forms in recent subaerial habitats and in Precambrian cherts: *Gloethece coerulea* Geitler and *Eosynechococcus moorei* Hoffmann. Precambrian Res 8:201-217

111. Golubic S, Hofmann HJ (1976) Comparison of modern and mid-Precambrian Entophysalidaceae (Cyanophyta) in stromatolitic algal mats: cell division and degradation. J Paleont 50:1074-1082

112. Golubic S, Seong-Joo L (1999) Early Cyanobacterial fossil record: preservation, palaeoenvironments and identification. Eur J Phycol 34:339-348

113. Gomes R, Levison HF, Tsiganis K, Morbidelli A (2005) Origin of the cataclysmic late heavy bombardment period of the terrestrial planets. Nature 435(7041):466-469

114. Gradzinski M, Chmiel MJ, Lewandowska A, Michalska-Kasperkiewicz B (2010) Siliciclastic micro-stromatolites in a sandstone cave: role of trapping and binding of detrital particles in formation of cave deposits. Ann Soc Geol Polon 80(3):303-314

115. Grandstaff DE, Edelman MJ, Foster RW, Zbinden E, Kimberley MM (1986) Chemistry and mineralogy of Precambrian paleosols at the base of the Dominion and Pongola Groups. Precambrian Res 32:97-131

116. Gray J, Boucot AJ (1994) Early Silurian nonmarine animal remains and the nature of the early continental ecosystem. Acta Palaeontol Polon 38(3–4):303-328

117. Gray J, Shear WA (1992) Early life on land. Am Sci 80:444-456

118. Gunatilaka A (1975) Some aspects of the biology and sedimentology of laminated algal mats from Mannar Lagoon, northwest Ceylon. Sediment Geol 14:275-300

119. Gutzmer J, Beukes NJ (1998) Earliest laterites and possible evidence for terrestrial vegetation in the Early Proterozoic. Geology 26:263-266

120. Hallbauer DK (1975) The plant origin of the Witwatersrand carbon. Minerals Sci Eng 7(2):111-131

121. Hallbauer DK, van Warmelo KT (1974) Fossilized plants in thucholite from Precambrian rocks of the Witwatersrand, South Africa. Precambrian Res 1:199-212

122. Hamblin WK, Christiansen EH (2007) Earth's dynamic systems. Upper Saddle River, NJ: Prentice Hall.

123. Han TM, Runnegar B (1992) Megascopic eukaryotic algae from the 2.1-billion-year-old negaunee iron-formation, Michigan. Science 257(5067):232-235

124. Hanel M, Montenari M, Kalt A (1999) Determining sedimentation ages of high-grade metamorphic gneisses by their palynological record: a case study in the northern Schwarzwald (Variscan Belt, Germany). Int J Earth Sci 88:49-59

125. Harrison SP, Kohfeld KE, Roelandt C, Claquin T (2001) The role of dust in climate changes today, at the last glacial maximum and in the future. Earth Sci Rev 54:43-80

126. Hartmann WK, Ryder G, Dones L, Grinspoon D (2000) The time-dependent intense bombardment of the primordial Earth/Moon system. In: Canup RM, Righter K (eds) Origin of the Earth and moon, Tucson: University of Arizona Press. pp 493-512

127. Hazen RM, Sverjensky DA (2010) Mineral surfaces, geochemical complexities, and the origins of life. In: Deamer D, Szostak JW (eds) The origins of life, volume 5, 2nd edn. Cold Spring Harbor: Cold Spring Harbor Laboratory Press. pp 1-21

128. Hazen RM, Golden J, Downs RT, Hystad G, Grew ES, Azzolini D, Sverjensky DA (2012) Mercury (Hg) mineral evolution: a mineralogical record of supercontinent assembly, changing ocean geochemistry, and the emerging terrestrial biosphere. Am Mineral 97:1013-1042

129. Heckman DS, Geiser DM, Eidell BR, Stauffer RL, Kardos NL, Hedges SB (2001) Molecular evidence for the early colonization of land by fungi and plants. Science 293(5532):1129-1133

130. Herrero A, Flores E (eds) (2008) The cyanobacteria: molecular biology, genomics and evolution Norfolk: Caister Academic Press.

131. Hickman AH, Van Kranendonk MJ (2012) Early Earth evolution: evidence from the 3.5-1.8 Ga geological history of the Pilbara region of Western Australia. Episodes 35(1):283-297

132. Hoashi M, Bevacqua DC, Otake T, Watanabe Y, Hickman AH, Utsunomiya S, Ohmoto H (2009) Primary haematite formation in an oxygenated sea 3.46 billion years ago. Nat Geosci 2:301-306

133. Hoffman PF (1995) The oldest terrestrial landscape. Nature 375(6532):537-538

134. Hoffman PF, Schrag DP (2002) The snowball Earth hypothesis: testing the limits of global change. Terra Nova 14:129-155

135. Hofmann HJ (1976) Precambrian microflora, Belcher Islands, Canada; significance and systematics. J Paleontol 50:1040-1073

136. Holland HD (1984) The chemical evolution of the atmosphere and oceans. Princeton: Princeton University Press.

137. Holland HD (1992) Distribution and paleoenvironment interpretation of Proterozoic paleosols. In: Schopf JW, Klein C (eds) The Proterozoic biosphere, Cambridge: Cambridge University Press. pp 153-155

138. Holm NG (1992) Marine hydrothermal systems and the origins of life. Orig Life Evol Biosph 22:181-242

139. Hoppert M, Reimer R, Kemmling A, Schröder A, Günzl B, Heinken T (2004) Structure and reactivity of a biological soil crust from a xeric sandy soil in central Europe. Geomicrobiol J 21:183-191

140. Horodyski RJ, Knauth PL (1994) Life on land in the Precambrian. Science 263(5146):494-498

141. Hu C, Liu Y, Song L, Zhang D (2002) Effect of desert soil algae on the stabilization of fine sands. J Appl Phycol 14(4):281-292

142. Hupe P (1952) Sur des problematica du Precambrien III. Division des Mines et de la Géologie, Service Géologique de Morocco, Notes et Memoires 103:297-383

143. Iizuka T, Horie K, Komiya T, Maruyama S, Hirata T, Hidaka H, Windley BF (2006) 4.2 Ga zircon xenocryst in an Acasta gneiss from northwestern Canada: evidence for early continental crust. Geology 34:245-248

144. Ishizuka H (2008) Protoliths of the Napier Complex in Enderby Land, East Antarctica; an overview and implication for crustal formation of Archean continents. J Miner Petrol Sci 103:218-225

145. Jackson TA (1967) Fossil Actinomycetes in Middle Precambrian glacial varves. Science, New Series 155(3765):1003-1005

146. Javaux EJ, Marshall CP, Bekker A (2010) Organic-walled microfossils in 3.2-billion-year-old shallow-marine siliciclastic deposits. Nature 463:934-938

147. Jickells TD, An ZS, Andersen KK, Baker AR, Bergametti G, Brooks N, Cao JJ, Boyd PW, Duce RA, Hunter KA, Kawahata H, Kubilay N, laRoche J, Liss PS, Mahowald N, Prospero JM, Ridgwell AJ, Tegen I, Torres R (2005) Global iron connections between desert dust, ocean biogeochemistry, and climate. Science 308:67-71

148. Johnson I, Watanabe Y, Stewart B, Ohmoto H (2009) Earth's oldest (~3.4 Ga) lateritic paleosol in the Pilbara Craton, Western Australia. Davos, Switzerland: Proceedings of the Goldschmidt conference.

149. Johnson I, Watanabe Y, Stewart B, Ohmoto H (2010) Evidence for terrestrial life and an O2-rich atmosphere in the oldest (~3.4 Ga) paleosol in the east Pilbara craton, Western Australia. League City, TX: Proceedings of the 6th Astrobiology Science Conference. 20–26 April 2010

150. Kandler O (1994) The early diversification of life. In: Bengston S (ed) Early life on earth, New York: Columbia University Press. pp 152-161

151. Kappler A, Pasquero C, Konhauser KO, Newman DK (2005) Deposition of banded iron formations by anoxygenic phototrophic Fe(II)-oxidizing bacteria. Geology 33(11):865-868

152. Kasting J (2009) How to find a habitable planet. Princeton: Princeton University Press.

153. Kato Y, Suzuki K, Nakamura K, Hickman AH, Nedachi M, Kusakabe M, Bevacqua DC, Ohmoto H (2009) Hematite formation by oxygenated groundwater more than 2.76 billion years ago. Earth Planet Sci Lett 278:40-49

154. Keller CK, Wood BD (1993) Possibility of chemical weathering before the advent of vascular land plants. Nature 364:223-225

155. Kendall B, Reinhard CT, Lyons TW, Kaufman AJ, Poulton SW, Anbar AD (2010) Pervasive oxygenation along late Archaean ocean margins. Nat Geosci 3(9):647-652

156. Kennedy MJ, Wagner T (2011) Clay mineral continental amplifier for marine carbon sequestration in a greenhouse ocean. Proc Nat Acad Sci USA 108:9776-9781

157. Kennedy MJ, Chadwick OA, Vitousek PM, Derry LA, Hendricks DM (1998) Replacement of weathering with atmospheric sources of base cations during ecosystem development, Hawaiian Islands. Geology 26:1015-1018

158. Kennedy M, Droser M, Mayer LM, Pevear D, Mrofka D (2006) Late Precambrian oxygenation; inception of the clay mineral factory. Science 311:1446-1449

159. Kimberley MM, Grandstaff DE (1986) Profiles of elemental concentrations in Precambrian paleosols on basaltic and granitic parent materials. Precambrian Res 32:133-154

160. Kirschvink JL, Kopp RE (2008) Palaeoproterozoic ice houses and the evolution of oxygen-mediating enzymes: the case for a late origin of photosystem II. Phil Trans R Soc B 363:2755-2765

161. Klein C, Beukes NJ, Schopf JW (1987) Filamentous microfossils in the early Proterozoic Transvaal supergroup: their morphology, significance, and paleoenvironmental setting. Precambrian Res 36:81-94

162. Knauth LP, Kennedy MJ (2009) The late Precambrian greening of the Earth. Nature 460:728-732

163. Knoll AH, Javaux EJ, Hewitt D, Cohen P (2006) Eukaryotic organisms in Proterozoic oceans. Philosophical Transactions of the Royal Society B 361(1470):1023-1038

164. Knoll A, Canfield D, Konhauser K (eds) (2012) Fundamentals of geobiology Chichester: Wiley-Blackwell.

165. Koehler I, Konhauser KO, Kappler A (2010) Role of microorganisms in banded iron formations. In: Barton L, Mandl M, Loy A (eds) Geomicrobiology: molecular and environmental perspective, New York: Springer. pp 309-324

166. Komiya T, Maruyama S, Nohda S, Masuda T, Hayashi M, Okamoto S (1999) Plate tectonics at 3.8–3.7 Ga; field evidence from the Isua accretionary complex, southern West Greenland. J Geol 107:515-554

167. Konhauser K (2007) Introduction to geomicrobiology. Oxford: Blackwell.

168. Konhauser KO, Schultzelam S, Ferris FG, Fyfe WS, Longstaffe FJ, Beveridge TJ (1994) Mineral precipitation by epilithic biofilms in the Speed River, Ontario, Canada. Appl Environ Microbiol 60:549-553

169. Konhauser KO, Newman DK, Kappler A (2005) The potential significance of microbial Fe(III) reduction during deposition of Precambrian banded iron formations. Geobiology 3:167-177

170. Kositcin N, Krapez B (2004) SHRIMP U-Pb detrital zircon geochronology of the late Archean Witwatersrand Basin, relation between zircon provenance age spectra and basin evolution. Precambrian Res 129:141-168

171. Labandeira CC (2005) Invasion of the continents: cyanobacterial crusts to tree-inhabiting arthropods. Trends Ecol Evol 20(5):253-262

172. Lalonde K, Mucci A, Ouellet A, Gélinas Y (2012) Preservation of organic matter in sediments promoted by iron. Nature 483(7388):198-200

173. Lannerbro R (1954) Description of some structures, possibly fossils, in Jotnian sandstone from Mångsbodarna in Dalecarlia. Geologiska Föreningens i Stockholm Förhandlingar 76:46-50

174. Lau KM, Kim MK, Kim KM (2006) Asian monsoon anomalies induced by aerosol direct effects. Clim Dyn 26:855-864

175. Lazcano A, Miller SL (1994) How long did it take for life to begin and evolve to cyanobacteria? J Mol Evol 39:546-554

176. Li W, Johnson CM, Beard BL (2012) U–Th–Pb isotope data indicate Phanerozoic age for oxidation of the 3.4 Ga Apex Basalt. Earth Planet Sci Lett 319–320:197-206

177. Liu YZ, Dong GR, Li CZ (1994) A study on the factors influencing soil erosion through wind tunnel experiments. Chin J Arid Land Res 7:359-367

178. Lopez-Garcia P, Moreira D, Douzery EJP, Forterre P, van Zuilen M (2006) Ancient fossil record and early evolution (ca. 3.8 to 0.5 Ga). In: Gargaud M (ed) From suns to life: a multidisciplinary approach to the history of life on Earth. Earth, moon, and planets, New York: Springer. pp 247-290

179. Lowe DR (1980) Stromatolites 3,400-Myr old from the Archean of Western Australia. Nature 284:441-443

180. Lozupone CA, Knight R (2007) Global patterns in bacterial diversity. Proc Natl Acad Sci USA 104:11436-11440

181. Macfarlane AW, Danielson A, Holland HD (1994) Geology and major and trace element chemistry of the late Archean weathering profiles in the Fortescue Group, Western Australia: implications for atmospheric pO_2. Precam Res 65:297-317

182. Madigan MT, Marrs BL (1997) Extremophiles. Scientific American 276(4):66-71

183. Madigan M, Martinko J, Parker J (2003) Brock: biology of microorganisms. Upper Saddle River, NJ: Pearson-Prentice Hall.

184. Malam Issa O, Le Bissonnais Y, Défarge C, Trichet J (2001) Role of a cyanobacterial cover on structural stability of sandy soils in the Sahalian part of western Niger. Geoderma 101:15-30

185. Martini JEJ (1994) A late Archaean–Palaeoproterozoic (2.6 Ga) palaeosol on ultramafics in the eastern Transvaal, South Africa. Precambrian Res 67:159-180

186. Mazor G, Kidron GJ, Vonshak A, Abeliovich A (1996) The role of cyanobacterial exopolysaccharides in structuring desert microbial crusts. FEMS Microbiol Ecol 21:121-130

187. McConnell RL (1974) Preliminary report of microstructures of probable biologic origin from the Mescal Formation (Proterozoic) of central Arizona. Precambrian Res 1(3):227-234

188. McKeegan KD, Kudryavtsev AB, Schopf JW (2007) Raman and ion microscopic imagery of graphitic inclusions in apatite from older than 3830 Ma Akilia supracrustal rocks, west Greenland. Geology 35:591-594

189. McTainsh GH, Strong CL (2007) The role of aeolian dust in ecosystems. Geomorphology 89(1–2):39-54

190. Meadow JF, Zabinski CA (2012) Spatial heterogeneity of eukaryotic microbial communities in an unstudied geothermal diatomaceous biological soil crust: Yellowstone National Park, WY, USA. FEMS Microbiol Ecol 82(1):182-191

191. Mloszewska AM, Pecoits E, Cates NL, Mojzsis SJ, O'Neil J, Robbins LJ, Konhauser KO (2012) The composition of Earth's oldest iron formations: the Nuvvuagittuq Supracrustal Belt (Quebec, Canada). Earth Planet Sci Lett 317–318:331-342

192. Mojzsis SJ, Harrison MT, Pidgeon RT (2001) Oxygen-isotope evidence from ancient zircons for liquid water at the Earth's surface 4,300 Myr ago. Nature 409:178-181

193. Mossman DJ, Minter WEL, Dutkiewicz A, Hallbauer DK, George SC, Hennigh Q, Reimer TO, Horscroft FD (2008) The indigenous origin of Witwatersrand "carbon". Precambrian Res 164:173-186

194. Mulkidjanian AY, Bychkovc AY, Dibrova DV, Galperin MY, Koonin EV (2012) Origin of the first cells at terrestrial, anoxic geothermal fields. Proc Natl Acad Sci USA 109:E821-E830

195. Myers JS (2001) Protoliths of the 3.8-3.7 Ga Isua Greenstone Belt, West Greenland. Precambrian Res 105:129-141

196. Myers JS (2004) Isua enigmas: illusive tectonic, sedimentary, volcanic and organic features of the >3.7 Ga Isua greenstone belt, southwest Greenland. In: Eriksson PG, Altermann W, Nelson DR, Mueller WU, Catuneanu O (eds) The Precambrian Earth: tempos and events. Developments in Precambrian geology, vol. 12, Amsterdam: Elsevier. pp 66-73

197. Nagy M, Perez A, Garcia-Pichel F (2005) The prokaryotic diversity of biological soil crusts in the Sonoran desert (Organ Pipe Cactus National Monument, AZ). FEMS Microbiol Ecol 54:233-245

198. Neher DA, Lewins SA, Weicht TR, Darby BJ (2009) Microarthropod communities associated with biological soil crusts in the Colorado Plateau and Chihuahuan deserts. J Arid Environ 73:672-677

199. Nelson DR (2004) Earth's formation and first billion years. In: Eriksson PG, Altermann W, Nelson DR, Mueller WU, Catuneanu O (eds) The Precambrian Earth: tempos and events,\. Developments in Precambrian geology, vol. 12, Amsterdam: Elsevier. pp 3-27

200. Nemergut DR, Costello EK, Hamady M, Lozupone C, Jiang L, Schmidt SK, Fierer N, Townsend AR, Cleveland CC, Stanish L, Knight R (2011) Global patterns in the biogeography of bacterial taxa. Environ Microbiol 13(1):135-144

201. Nesbitt HW, Young GM (2004) Aggressive Archaean weathering. In: Eriksson PG, Altermann W, Nelson DR, Mueller WU, Catuneanu O (eds) The Precambrian Earth: tempos and events. Developments in Precambrian geology, vol. 12, Amsterdam: Elsevier. pp 482-493

202. Nicholas J, Bildgen P (1979) Relations between the location of the karst bauxites in the northern hemisphere, the global tectonics, and the climatic variations during geological time. Palaeogeogr Palaeoclimateol Palaeoecol 28:205-239

203. Noffke N (2010) Geobiology: microbial mats in sandy deposits from the Archean era to today. Berlin: Springer.

204. Noffke N (2011) A Modern Perspective on Ancient Life: Microbial Mats in Sandy Marine Settings from the Archean Era to Today. In: Golding SD, Glikson M (eds) Earliest Life on Earth: Habitats. Environments and Methods of Detection, Dordrecht: Springer. pp 171-182

205. Noffke N, Gerdes G, Klenke T, Krumbein WE (2001) Perspectives. Microbially induced sedimentary structures—a new category within the classification of primary sedimentary structures. J Sediment Res 71(5):649-656

206. Noffke N, Eriksson KA, Hazen RM, Simpson EL (2006) A new window into Early Archean life: microbial mats in Earth's oldest siliciclastic tidal deposits (3.2 Ga Moodies Group, South Africa). Geology 34(4):253-256

207. Noffke N, Beukes N, Bower D, Hazen RM, Swift DJP (2008) An actualistic perspective into Archean worlds – (cyano-)bacterially induced sedimentary structures in the siliciclastic Nhlazatse Section, 2.9 Ga Pongola Supergroup, South Africa. Geobiology 6:5-20

208. Noffke N, Christian DR, Hazen RM (2011) A (cyano-)bacterial ecosystem in the Archean 3.49 Ga Dresser Formation, Pilbara, Western Australia. GSA Annual Meeting in Minneapolis. Paper No. 56–11. Geological Society of America Abstracts with Programs 43(5):159

209. Nutman AP (2006) Antiquity of the oceans and continents. Elements 2:223-227

210. Nutman AP, McGregor VR, Friend CRL, Bennett VC, Kinny PD (1996) The Itsaq Gneiss Complex of southern West Greenland; the world's most extensive record of early crustal evolution (3900–3600 Ma). Precambrian Res 78:1-39

211. Nutman AP, Friend CRL, Bennett VC (2002) Evidence for 3650–3600 Ma assembly of the northern end of the Itsaq Gneiss

Complex, Greenland: implication for early Archaean tectonics. Tectonics 21(1):5-1–5-28

212. Nutman AP, Friend CRL, Bennett VC, Wright D, Norman MD (2010) ≥3700 Ma pre-metamorphic dolomite formed by microbial mediation in the Isua supracrustal belt (W. Greenland): simple evidence for early life? Precambrian Res 183:725-737

213. O'Neil J, Francis D, Carlson RW (2011) Implications of the Nuvvuagittuq greenstone belt for the formation of earth's early crust. J Petrol 52:985-1009

214. Och LM, Shields-Zhou GA (2012) The Neoproterozoic oxygenation event. Environmental perturbations and biogeochemical cycling. Earth Sci Rev 110(1–4):26-57

215. Ohmoto H (1996) Evidence in pre-2.2 Ga paleosols for the early evolution of atmospheric oxygen and terrestrial biota. Geology 24:1135-1138

216. Ohmoto H (2004) Archean atmosphere, hydrosphere, and biosphere. In: Eriksson PG, Altermann W, Nelson DR, Mueller WU, Catuneanu O (eds) The Precambrian Earth: tempos and events. Developments in Precambrian geology, vol. 12, Amsterdam: Elsevier. pp 361-368

217. Ohmoto H, Watanabe Y, Allwood A, Burch I, Knauth P, Yamaguchi K, Johnson I, Altinok E (2007) Formation of probable lateritic soils ~3.43 Ga in the Pilbara Craton, Western Australia. Geochimica et Cosmochimica Acta, Supplement 71(15):A733

218. Omelon CR, Pollard WH, Ferris FG (2006) Chemical and ultrastructural characterization of high arctic cryptoendolithic habitats. Geomicrobiol J 23:189-200

219. O'Neil J, Carlson RW, Francis D, Stevenson RK (2009) Response to comment on "Neodymium-142 evidence for Hadean mafic crust". Science 325:267b

220. Pace N (1997) A molecular view of microbial diversity and the biosphere. Science 276:734-740

221. Palmer JA, Phillips GN, McCarthy TS (1989) Paleosols and their relevance to Precambrian atmospheric composition. J Geol 97:77-92

222. Pandit MK, de Wall H, Chauhan NK (2008) Paleosol at the Archean-Proterozoic contact in NW India revisited: evidence for

oxidizing conditions during paleo-weathering? J Earth Syst Sci 117(3):201-209

223. Papineau D, DeGregorio BT, Cody GD, Fries MD, Mojzsis SJ, Steele A, Stroud RM, Fogel ML (2010) Ancient graphite in the Eoarchean quartz-pyroxene rock from Akilia in southwest Greenland I: petrographic and spectroscopic characterization. Geochim Cosmochim Acta 74:5862-5883

224. Papineau D, DeGregorio BT, Stroud RM, Steele A, Pecoits E, Konhauser K, Wang J, Fogel ML (2010) Ancient graphite in the Eoarchean quartz-pyroxene rock from Akilia in southern West Greenland II: isotopic and chemical compositions and comparison with Paleoproterozoic banded iron formations. Geochim Cosmochim Acta 74:5884-5905

225. Papineau D, De Gregorio BT, Cody GD, O'Neil J, Steele A, Stroud RM, Fogel ML (2011) Young poorly crystalline graphite in the 3.8-Gyr-old Nuvvuagittuq banded iron formation. Nat Geosci 4(6):376-379

226. Paul EA, Collins HP, Leavitt SW (2001) Dynamics of resistant soil carbon of Midwestern agricultural soils measured by naturally occurring ^{14}C abundance. Geoderma 104(3–4):239-256

227. Porter ML, Engel AS, Kinkle B, Kane TC (2009) Productivity-diversity relationships from chemolithoautotrophically based sulfidic karst systems. Int J Speleol 38:27-40

228. Potts M (1999) Mechanisms of desiccation tolerance in cyanobacteria. Eur J Phycol 34:319-328

229. Potts M, Friedmann EI (1981) Effects of water stress on cryptoendolithic cyanobacteria from hot desert rocks. Arch Microbiol 130:267-271

230. Prasad N, Roscoe SM (1996) Evidence of anoxic to oxic atmospheric change during 2.45-2.22 Ga from lower and upper sub-Huronian paleosols, Canada. Catena 27:105-121

231. Prave AR (2002) Life on land in the Proterozoic: evidence from the Torridonian rocks of northwest Scotland. Geology 30(9):811-814

232. Pufahl PK, Hiatt EE (2012) Oxygenation of the Earth's atmosphere–ocean system: a review of physical and chemical sedimentologic responses. J Mar Petrol Geol 32(1):1-20

233. Rasmussen B (2000) Filamentous microfossils in a 3,235-million-year-old volcanogenic massive sulphide deposit. Nature 405:676-679

234. Rasmussen B, Fletcher IR, Brocks JJ, Kilburn MR (2008) Reassessing the first appearance of eukaryotes and cyanobacteria. Nature 455(7216):1101-1104

235. Rasmussen B, Blake TS, Fletcher IR, Kilburn MR (2009) Evidence for microbial life in synsedimentary cavities from 2.75 Ga terrestrial environments. Geology 37:423-426

236. Reddy SG, Garcia-Pichel F (2006) The community and phylogenetic diversity of biological soil crusts in the Colorado Plateau studied by molecular fingerprinting and intensive cultivation. Microb Ecol 52:345-357

237. Reimer TO (1986) Alumina-rich rocks from the Early Precambrian of the Kaapvaal craton as indicators of paleosols and as products of other decompositional reactions. Precambrian Res 32:155-179

238. Retallack GJ (1986) The fossil record of soils. In: Wright PV (ed) Paleosols: their recognition and interpretation, Oxford: Blackwell. pp 1-57

239. Retallack GJ (1986) Reappraisal of a 2200 Ma-old paleosol near Waterval Onder, South Africa. Precambrian Res 32:195-232

240. Retallack GJ (2001) Soils of the past. London: Blackwell Science.

241. Retallack GJ (2009) Cambrian–Ordovician non-marine fossils from South Australia. Alcheringa 33:355-391

242. Retallack (2001) Criteria for distinguishing microbial mats and earths. In: Noffke N, Chafetz H (ed) Microbial mats in siliciclastic depositional systems through time. SEPM Special Publication no. 101. Society for Sedimentary Geology, Tulsa, OK, pp. 139-152

243. Retallack GJ (2013) Ediacaran life on land. Nature 493(7430):89-92

244. Retallack GJ, Mindszenty A (1994) Well preserved Late Precambrian paleosols from northwest Scotland. J Sediment Res A64:264-281

245. Reynolds R, Belnap J, Reheis M, Lamothe P, Luiszer F (2001) Aeolian dust in Colorado Plateau soils: nutrient inputs and recent change in source. Proc Natl Acad Sci USA 98:7123-7127

246. Rino S, Komiya T, Windley BF, Katayama S, Motoki A, Hirata T (2004) Major episodic increases of continental crust growth determined from zircon ages of river sands; implication for mantle overturns in the early Precambrian. Phys Earth Planet Inter 146:369-394

247. Romans BW, Graham SA (2013) A deep-time perspective of land-ocean linkages in the sedimentary record. Annu Rev Mar Sci 5:69-94

248. Rosing MT, Frei R (2004) U-rich Archaean sea-floor sediments from Greenland: indications of > 3700 Ma oxygenic photosynthesis. Earth Planet Sc Lett 217:237-244

249. Rosing MT, Bird DK, Sleep NH, Glassley W, Albarede F (2006) The rise of continents – an essay on the geologic consequences of photosynthesis. Palaeogeogr Palaeoclimateol Palaeoecol 232:99-113

250. Rossi F, Potrafka RM, Garcia-Pichel F, De Philippis R (2012) The role of the exopolysaccharides in enhancing hydraulic conductivity of biological soil crusts. Soil Biol Biochem 46:33-40

251. Rye R, Holland HD (2000) Life associated with a 2.76 Ga ephemeral pond? Evidence from Mount Roe #2 paleosol. Geology 28:483-486

252. Santosh M (2010) A synopsis of recent conceptual models on supercontinent tectonics in relation to mantle dynamics, life evolution and surface environment. J Geodyn 50(3–4):116-133

253. Sarbu SM, Kane TC, Kinkle BK (1996) A chemoautotrophically based cave ecosystem. Science 272:1953-1955

254. Schau M, Henderson JB (1983) Archean chemical weathering at three localities on the Canadian shield. Dev Precambrian Geol 7:81-116

255. Schieber J, Bose PK, Eriksson PG, Banerjee S, Sarkar S, Altermann W, Catuneau O (eds) (2007) Atlas of microbial mat features preserved within the siliclastic rock record Amsterdam: Elsevier.

256. Schiffbauer JD, Xiao S (2009) Novel application of focused ion beam-electron microscopy (FIB-EM) in preparation and analysis of microfossils ultrastructures: a new view of complexity in early eukaryotic organisms. Palaios 24:616-626

257. Schiffbauer JD, Yin L, Bodnar RJ, Kaufman AJ, Meng F, Hu J, Shen B, Yuan X, Bao H, Xiao S (2007) Ultrastructural and geochemical characterization of Archean-Paleoproterozoic graphite particles: implications for recognizing traces of life in highly metamorphosed rocks. Astrobiology 7:684-704

258. Schiffbauer JD, Wallace AF, Hunter JL Jr, Kowalewski M, Bodnar RJ, Xiao S (2012) Thermally-induced structural and chemical alteration of organic-walled microfossils: an experimental approach to understanding fossil preservation in metasediments. Geobiology 10(5):402-423

259. Schirrmeister BE, de Vosb JM, Antonelli A, Bagheri HC (2013) Evolution of multicellularity coincided with increased diversification of cyanobacteria and the Great Oxidation Event. PNAS-USA

260. Schmidt SK, Reed SC, Nemergut DR, Stuart-Grandy A, Cleveland CC, Weintraub MN, Hill AW, Costello EK, Meyer AF, Neff JC, Martin AM (2008) The earliest stages of ecosystem succession in high-elevation (5000 metres above sea level), recently deglaciated soils. Proc Biol Sci 275(1653):2793-2802

261. Schopf JW (1968) Microflora of the bitter springs formation, late Precambrian, central Australia. J Paleontol 42:651-688

262. Schopf JW (ed) (1983) Earth's earliest biosphere Princeton, NJ: Princeton University Press.

263. Schopf JW (1992) Paleobiology of the Archean. In: Schopf JW, Klein C (eds) The Proterozoic biosphere, New York: Cambridge University Press.

264. Schopf JW (1992) Proterozoic prokaryotes: affinities, geologic distribution, and evolutionary trends. In: Schopf JW, Klein C (eds) The Proterozoic biosphere, New York: Cambridge University Press.

265. Schopf JW, Klein C (eds) (1992) The Proterozoic biosphere New York: Cambridge University Press.

266. Schopf JW, Walter MR, Ruiji C (2007) Earliest evidence of life on earth. Precambrian Res 158:139-140

267. Schwartzman DW, Volk T (1989) Biotic enhancement of weathering and the habitability of earth. Nature 340:457-460

268. Shear WA (1991) The early development of terrestrial ecosystems. Nature 351:283-289

269. Sheldon ND (2012) Microbially induced sedimentary structures in the ca. 1100 Ma terrestrial midcontinent rift of North America. In: Noffke N, Chafetz H (eds) Microbial mats in siliciclastic depositional systems through time. SEPM Special Publication no. 101, Tulsa, OK: Society for Sedimentary Geology. pp 153-162

270. Shen Y, Buick R (2004) The antiquity of microbial sulfate reduction. Earth Sci Rev 64:243-272

271. Shen Y, Farquhar J, Masterson A, Kaufman AJ, Buick R (2009) Evaluating the role of microbial sulfate reduction in the early Archean using quadruple isotope systematics. Earth Planet Sc Lett 279:383-391

272. Shephard KL (1987) Evaporation of water from the mucilage of a gelatinous algal community. Br Phycol J 22:181-185

273. Sigler WV, Crivii S, Zeyer J (2002) Bacterial succession in glacial forefield soils characterized by community structure, activity and opportunistic growth dynamics. Microb Ecol 44:306-316

274. Simpson WS, Simpson EL, Wizevich MC, Malenda HF, Hilbert-Wolf HL, Tindall SE (2010) A preserved Late Cretaceous biological soil crust in the capping sandstone member, Wahweap Formation, Grand Staircase-Escalante National Monument, Utah, Palaeoclimatic implications. Sediment Geol 230:139-145

275. Singh SP, Kumari S, Rastogi RP, Singh KL, Richa SRP (2010) Photoprotective and biotechnological potentials of cyanobacterial sheath pigment, scytonemin. Afr J Biotechnol 9:580-588

276. Sinha RP, Häder DP (2002) UV-induced DNA damage and repair: a review. Photochem Photobiol Sci 1:225-236

277. Skotnicki SJ, Knauth LP (2007) The Middle Proterozoic Mescal Paleokarst, Central Arizona, USA: karst development, silicification, and cave deposits. J Sediment Res 77(12):1046-1062

278. Sleep NH, Bird DK (2007) Niches of the pre-photosynthetic biosphere and geologic preservation of Earth's earliest ecology. Geobiology 5:101-117

279. Smith SE, Read D (2008) Mycorrhizal symbiosis. New York: Elsevier.

280. Squire RJ, Stewart IR, Zang WL (2006) Acritarchs in polydeformed and highly altered Cambrian rocks in western Victoria. Aust J Earth Sci 53:697-705

281. Stern RJ, Scholl DW (2010) Yin and yang of continental crust creation and destruction by plate tectonic processes. Int Geol Rev 52:1-31

282. Strother PK, Battison L, Brasier MD, Wellman CH (2011) Earth's earliest non-marine eukaryotes. Nature 473:505-509

283. Stüeken EE, Catling DC, Buick R (2012) Contributions to late Archaean sulphur cycling by life on land. Nat Geosci 5:722-725

284. Summons RE, Jahnke LL, Hope JM, Logan GA (1999) 2-Methylhopanoids as biomarkers for cyanobacterial oxygenic photosynthesis. Nature 400:554-557

285. Sun HJ, Friedmann EI (1999) Growth on geological time scales in the Antarctic cryptoendolithic microbial community. Geomicrobiol J 16:193-202

286. Takeuchi N, Kohshima S, Seko K (2001) Structure, formation, darkening process of albedo reducing material (cryoconite) on a Himalayan glacier: a granular algal mat growing on the glacier. Arct Antarct Alp Res 33:115-122

287. Taylor TN, Taylor EL, Krings M (2009) Paleobotany: the biology and evolution of fossil plants. Amsterdam: Elsevier.

288. Thiry M, Simon-Coincon R (eds) (1999) Palaeoweathering, palaeosurfaces and related continental deposits. Special publication 27 of the International Association of Sedimentologists Oxford: Blackwell Science.

289. Tice MM, Lowe DR (2004) Photosynthetic microbial mats in the 3,416-Myr-old ocean. Nature 431:549-552

290. Tirkey J, Adhikary SP (2005) Cyanobacteria in the biological soil crusts of India. Curr Sci 89:515-521

291. Tomitani A, Knoll AH, Cavanaugh CM, Ohno T (2006) The evolutionary diversification of cyanobacteria: molecular-phylogenetic and paleontological perspectives. Proc Natl Acad Sci USA 103:5442-5447

292. Trendall AF, Blockley JG (2004) Precambrian iron-formations. In: Eriksson PG, Altermann W, Nelson DR, Mueller WU, Catuneanu O (eds) The Precambrian Earth: tempos and events. Developments

in Precambrian geology, vol. 12, Amsterdam: Elsevier. pp 403-420

293. Ueno Y, Yamada K, Yoshida N, Maruyama S, Isozaki Y (2006) Evidence from fluid inclusions for microbial methanogenesis in the early Archaean era. Nature 440:516-519

294. Van Kranendonk MJ (2004) Archaean tectonics 2004: a review. Precambrian Res 131:143-151

295. Van Kranendonk MJ (2011) Stromatolite morphology as an indicator of biogenicity for Earth's oldest fossils from the 3.5-3.4 Ga Pilbara Craton, Western Australia. In: Reitner J, Queric NV, Arp G (eds) Advances in stromatolite geobiology. Lecture Notes in Earth Sciences, vol. 131, Germany: Springer.

296. Kranendonk MJ Van, Smithies RH, Bennett V (eds) (2007a) Earth's oldest rocks. Developments in Precambrian geology, series 15 Amsterdam: Elsevier.

297. Van Kranendonk MJ, Smithies RH, Hickman AH, Champion DC (2007) Review: secular tectonic evolution of Archean continental crust: interplay between horizontal and vertical processes in the formation of the Pilbara Craton, Australia. Terra Nova 19(1):1-38

298. Van Kranendonk MJ, Philippot P, Lepot K, Bodorkos S, Pirajno F (2008) Geological setting of Earth's oldest fossils in the c. 3.5 Ga Dresser Formation, Pilbara Craton, Western Australia. Precambrian Res 167:93-124

299. van Zuilen MA, Lepland A, Arrhenius G (2002) Reassessing the evidence for the earliest traces of life. Nature 418:627-630

300. Voigt E (1972) Tonrollen als potentielle Pseudofossilien. Nat Mus 102(11):401-410

301. Wacey D, Kilburn MR, Saunders M, Cliff J, Brasier MD (2011) Microfossils of sulphur-metabolizing cells in 3.4-billion-year-old rocks of Western Australia. Nat Geosci 4:698-702

302. Walsh MM (1992) Microfossils and possible microfossils from the Early Archean Onverwacht Group, Barberton Mountain Land, South Africa. Precambrian Res 54:271-292

303. Walsh MM, Lowe DR (1985) Filamentous microfossils from the 3500 Myr-old Onverwacht Group, Barberton Mountain Land, South Africa. Nature 314:530-532

304. Walsh MM, Lowe DR (1999) Modes of accumulation of carbonaceous matter in the early Archaean: A petrographic and geochemical study of the carbonaceous cherts of the Swaziland Supergroup. In: Lowe DR, Byerly GR (eds) Geologic Evolution of the Barberton Greenstone Belt, South Africa, Geological Society of America Special Paper 329, pp. 115-132

305. Walter MR (1983) Archean stromatolites: evidence of the Earth's earliest benthos. In: Schopf JW (ed) Earth's earliest biosphere, Princeton: Princeton University Press. pp 187-213

306. Walter MR, Buick R, Dunlop JSR (1980) Stromatolites 3,400–3,500 Myr old from the North Pole area, Western Australia. Nature 284:443-445

307. Watanabe Y, Martini JEJ, Ohmoto H (2000) Geochemical evidence for terrestrial ecosystems 2.6 billion years ago. Nature 408:574-578

308. Welch SA, Barker WW, Banfield JF (1999) Microbial extracellular polysaccharides and plagioclase dissolution. Geochim Cosmochim Acta 63:1405-1419

309. Westall F (2010) Early life: nature, distribution and evolution. In: Gargaud M, López-Garcìa P, Martin H (eds) Origins and evolution of life, Cambridge: An astrobiological perspective. Cambridge University Press. pp 391-413

310. Westall F, De Wit MJ, Dann J, Van Der Gaast S, De Ronde C, Gerneke D (2001) Early Archaean fossil bacteria and biofilms in hydrothermally influenced, shallow water sediments, Barberton Greenstone Belt, South Africa. Precambrian Res 106:91-112

311. Westall F, de Vries ST, Nijman W, Rouchon V, Orberger B, Pearson V, Watson J, Verchovsky A, Wright I, Rouzaud JN, Marchesini D, Anne S (2006a) The 3.466 Ga Kitty's Gap Chert, an Early Archaean microbial ecosystem. In: Reimold WU, Gibson R (eds) Processes on the early Earth. Geological Society of America special paper 405, Boulder, CO: Geological Society of America. pp 105-131

312. Westall F, de Ronde CEJ, Southam G, Grassineau N, Colas M, Cockell C, Lammer H (2006) Implications of a 3.472–3.333 Ga-old subaerial microbial mat from the Barberton Greenstone Belt, South Africa for the UV environmental conditions on the early Earth. Phil Trans R Soc B 361:1857-1875

313. Westall F, Cavalazzi B, Lemelle L, Marrocchi Y, Rouzaud JN, Simionovici A, Salomé M, Mostefaoui S, Andreazza C, Foucher F, Toporski J, Jauss A, Thiel V, Southam G, MacLean L, Wirick S, Hofmann A, Meibom A, Robert F, Défarge C (2011) Implications of in situ calcification for photosynthesis in a 3.3 Ga-old microbial biofilm from the Barberton greenstone belt, South Africa. Earth Planet Sci Lett 310(3–4):468-479

314. White D (2000) the physiology and biochemistry of prokaryotes. Oxford: Oxford University Press.

315. Whitton BA, Potts M (2000) the ecology of cyanobacteria: their diversity in time and space. Dordrecht: Kluwer.

316. Wilde SA, Valley JW, Peck WH, Graham CM (2001) Evidence from detrital zircons for the existence of continental crust and oceans on the Earth 4.4 Gyr ago. Nature 409:175-178

317. Williams JD, Dobrowolsk JP, West NE, Gillette DA (1995) Microphytic crust influence on wind erosion. Trans ASAE 38:131-137

318. Williams AJ, Buck BJ, Beyene MA (2012) Biological soil crusts in the Mojave Desert, USA: micromorphology and pedogenesis. Soil Sci Soc Am J 76(5):1685-1695

319. Windley B (2007) Overview and history of investigation of early Earth rocks. In: Van Kranendonk MJ, Smithies RH, Bennett V (eds) Earth's oldest rocks. Developments in Precambrian geology, series 15, Amsterdam: Elsevier. pp 3-7

320. Wright VP (1985) The precursor environment for vascular plant colonization. Phil Trans R Soc London B 309:143-145

321. Xiao S, Kaufman AJ (eds) (2006) Neoproterozoic geobiology and paleobiology. Topics in Geobiology, vol. 27 Dordrecht: Springer.

322. Xiao S, Knauth LP (2013) Palaeontology: fossils come in to land. Nature 493(7430):28-29

323. Xiong J, Fischer WM, Inoue K, Nakahara M, Bauer CE (2000) Molecular evidence for the early evolution of photosynthesis. Science 289:1724-1730

324. Yang W, Holland HD (2003) The Hekpoort paleosol in strata 1 at Garborone, Botswana: soil formation during the Great Oxidation Event. Am J Sci 303:187-220

325. Yasui A, McCready SJ (1998) Alternative repair pathways for UV-induced DNA damage. Bioessays 20(4):291-297

326. Yu G, Jacobsen SB (2011) Fast accretion of the Earth with a late Moon-forming giant impact. PNAS USA 108(43):17604-9

327. Zang WL (2007) Deposition and deformation of late Archean sediments and preservation of microfossils in the Harris Greenstone Domain, Gawler Craton, South Australia. Precambrian Res 156:107-124

328. Zbinden EA, Holland HD, Feakes CR, Dobos SK (1988) The Sturgeon Falls paleosol and the composition of the atmosphere 1.1 Ga BP. Precambrian Res 42:141-163

329. Zhang J (1992) Observation on algal effects on subaerial karst sedimentation. Geogr Res 11(2):26-33

330. Zhao B, Robb LJ, Harris C, Jordaan LJ (2006) Origin of hydrothermal fluids and gold mineralization associated with the Ventersdorp contact reef, Witwatersrand Basin, South Africa: constraints from S, O, and H isotopes. In: Reimold WU, Gibson RL (eds) Processes on the early Earth. Geological Society of America special paper 405, Boulder, CO: Geological Society of America. pp 333-352

331. Zhuravlev AY, Riding R (eds) (2001) The ecology of the Cambrian radiation New York: Columbia University Press.

332. Zonneveld KAF, Versteegh GJM, Kasten S, Eglinton TI, Emeis KC, Huguet C, Koch BP, De Lange GJ, De Leeuw JW, Middelburg JJ, Mollenhauer G, Prahl FG, Rethemeyer J, Wakeham SG (2010) Selective preservation of organic matter in marine environments; processes and impact on the sedimentary record. Biogeosciences 7:483-511

Citations

CHAPTER 1

Tomohiro Hajima, Michio Kawamiya, Michio Watanabe, Etsushi Kato, Kaoru Tachiiri, Masahiro Sugiyama, Shingo Watanabe, Hideki Okajima, and Akinori Ito, Modeling in Earth System Science up to and beyond IPCC AR5, doi: 10.1186/s40645-014-0029-y.

CHAPTER 2

Den Volokin and Lark ReLlez, on the average temperature of airless spherical bodies and the magnitude of Earth's atmospheric thermal effect, doi: 10.1186/2193-1801-3-723.

CHAPTER 3

Masaki Satoh, Hirofumi Tomita, Hisashi Yashiro, Hiroaki Miura, Chihiro Kodama, Tatsuya Seiki, Akira T Noda, Yohei Yamada, Daisuke Goto, Masahiro Sawada, Takemasa Miyoshi, Yosuke Niwa, Masayuki Hara, Tomoki Ohno, Shin-ichi Iga, Takashi Arakawa, Takahiro Inoue and Hiroyasu Kubokawa, The Non-Hydrostatic Icosahedral Atmospheric Model: Description and Development, doi:10.1186/s40645-014-0018-1.

CHAPTER 4

Jason W Barnes, Christophe Sotin, Jason M Soderblom, Robert H Brown, Alexander G Hayes, Mark Donelan, Sebastien Rodriguez, Stéphane Le Mouélic, Kevin H Baines, and Thomas B McCord, *Cassini/VIMS* Observes Rough Surfaces on Titan's Punga Mare in Specular Reflection, doi:10.1186/s13535-014-0003-4.

CHAPTER 5

Hugo Beraldi-Campesi, Early Life on Land and the First Terrestrial Ecosystems, doi: 10.1186/2192-1709-2-1.

Index